WELDING LICENSING EXAM
STUDY GUIDE

About the Authors

Rex Miller is Professor Emeritus of Industrial Technology at the State University College at Buffalo. He has taught technical curriculums at the high school, technical school, and college levels for more than 40 years. Dr. Miller is the author or coauthor of over 100 textbooks for vocational and industrial arts programs, including McGraw-Hill's best-selling *Carpentry and Construction* and *Electrician's Pocket Manual.*

Mark R. Miller is Professor and Coordinator of Industrial Technology at the University of Texas at Tyler. He has taught technical curriculums at the high school, technical school, and college levels for more than 20 years. Dr. Miller is the author or coauthor of over 30 textbooks for trade and technical programs, including *Audel Refrigeration: Home and Commercial* and *Audel Air Conditioning: Home and Commercial.*

WELDING LICENSING EXAM
STUDY GUIDE

Rex Miller
Mark R. Miller

New York   Chicago   San Francisco
Lisbon   London   Madrid   Mexico City
Milan   New Delhi   San Juan
Seoul   Singapore   Sydney   Toronto

The McGraw·Hill Companies

Library of Congress Cataloging-in-Publication Data

Miller, Rex, 1929-
    Welding licensing exam study guide / Rex Miller, Mark R. Miller.
        p. cm.
    ISBN 0-07-149376-X (alk. paper)
    1. Welding—Examinations—Study guides. 2. Welding—Examinations, questions, etc. 3. Welders (Persons)—Certification. I. Miller, Mark R. II. Title.
TS227.55.M55 2007
671.5'2076—dc22

                    2007011117

McGraw-Hill books are available at special quantity discounts to use as premiums and sales promotions, or for use in corporate training programs. For more information, please write to the Director of Special Sales, Professional Publishing, McGraw-Hill, Two Penn Plaza, New York, NY 10121-2298. Or contact your local bookstore.

Welding Licensing Exam Study Guide

2 3 4 5 6 7 8 9 10 IBT/IBT 1 9 8 7 6 5 4 3 2 1

ISBN-13: 978-0-07-149376-5

ISBN-10:   0-07-149376-X

Sponsoring Editor
LARRY HAGER

Editorial Supervisor
JODY MCKENZIE

Project Manager
VIRGINIA HOWE

Acquisitions Coordinator
LAURA HAHN

Copy Editor
WENDY LOCHNER

Proofreader
ROGER WOODSON

Production Supervisor
JIM KUSSOW

Composition
LONE WOLF ENTERPRISES, LTD.

Illustration
LONE WOLF ENTERPRISES, LTD.

This book is dedicated to

Patricia Ann Navara Miller

wife and mother

Contents

PART A **WELDING SAFETY**

PART B **METALS CUTTING**

---

* Space limitations preclude our including text material covering all the questions in each chapter. Outside sources should be consulted as you examine the whole field of welding.

PART E   **IRONS**

PART F   STEELS

PART G   ALLOYS

PART H   JOINTS AND WELDS

Preface

This book has been designed for use by welders and apprentices who work with the welding of metals of all descriptions. We assume that the reader has a working knowledge of basic welding and metallurgy. The book builds on this base, refreshes for some and introduces a wider range of topics for others.

The primary function of the question and answer format is to make it possible to develop your reasoning ability and organize your thoughts and knowledge in this very important field. Inasmuch as some of the work the welder does requires detailed instructions and skills, it is necessary for those who do this work to be able to demonstrate their practical knowledge and their ability to read, write, and think through the problems handed to them in everyday situations at work.

There are about 1,000 questions presented here, each designed to aid in the knowledge base needed to pass written exams in the specialty areas and become licensed to do the work accurately and skillfully.

Whether it's a need to brush up on the material or to improve your test-taking skills the book should aid you in your processing the information.

There are True/False questions, multiple-choice questions, and some matching type problems. All these questions have answers printed for easy access and learning.

The text material shown in the introduction to each chapter is there to aid in refreshing your memory and re-enforcing you basic knowledge of the many processes and procedures. Not every question is answered in the text material. It would take a stack of books to cover all the topics.

As you know, it is not possible to learn to swim without first getting a feel for the water. Neither is it possible to learn all there is to know about welding and metals by reading about them. You must be willing to get practical experience and devote time to the development of skills related to the job.

Obviously, not all problems can be present here; a great deal of on-the-job ingenuity is required.

It is the authors' hope that the book will serve as an appetizer to those really interested in going into this exciting, yet demanding field and help those who want to pass the certification exam for advancement in the trade.

Rex Miller
Mark R. Miller

Acknowledgments

No author works without being influenced and aided by others. Every book reflects this fact. A number of people cooperated in providing technical data and illustrations. For this we are grateful.

We would like to thank James Brumbaugh, Don Geary, and others who have contributed so much to the welding trade through their works. We would like to thank those organizations that so generously contributed information and illustrations.

Those who have been very helpful and cooperative are listed below:

American Welding Society
Bastian-Blessing Co.
Lincoln Electric Co.
Linde Div. of Union Carbide
Miller Welding Co.
Turner Div. of Clean-weld Products, Inc.
Victor Equipment Co.

Introduction

Welding, Soldering, and Brazing Workers

More than 6 out of 10 jobs are found in manufacturing industries. Training ranges from a few weeks of school or on-the-job training for low-skilled positions to several years of combined school and on-the-job training for highly skilled jobs. Employment is projected to grow more slowly than average. Job prospects should be excellent as employers report difficulty finding enough qualified people.

NATURE OF THE WORK

Welding is the most common way of permanently joining metal parts. In this process, heat is applied to metal pieces, melting and fusing them to form a permanent bond. Because of its strength, welding is used in shipbuilding, automobile manufacturing and repair, aerospace applications, and thousands of other manufacturing activities. Welding also is used to join beams when constructing buildings, bridges, and other structures, and to join pipes in pipelines, power plants, and refineries.

Welders use many types of welding equipment set up in a variety of positions, such as flat, vertical, horizontal, and overhead. They may perform manual welding, in which the work is entirely controlled by the welder, or semiautomatic welding, in which the welder uses machinery, such as a wire feeder, to perform welding tasks.

There are about 100 different types of welding. Arc welding is the most common type. Standard arc welding involves two large metal alligator clips that carry a high electrical current. One clip is attached to any part of the work piece being welded. The second clip is connected to a thin welding rod. When the rod touches the work piece, a powerful electrical circuit flows.

The massive heat created by the electrical arc causes both the work piece and the steel core of the rod to melt and run together. The hot metal cools quickly to form a solid bond. During welding, the flux that surrounds the rod's core vaporizes, forming an inert gas that serves to protect the weld from atmospheric elements that might weaken it; consequently, welding speed is important. Variations in speed can change the amount of flux applied, weakening the weld, or weakening the surrounding metal by increasing heat exposure.

Two common but advanced types of arc welding are Tungsten Inert Gas (TIG) and Metal Inert Gas

(MIG) welding. TIG welding often is used with stainless steel or aluminum. While TIG uses welding rods, MIG uses a spool of continuously fed wire, which allows the welder to join longer stretches of metal without stopping to replace the rod. In TIG welding, the welder holds the welding rod in one hand and an electric torch in the other hand. The torch is used to simultaneously melt the rod and the work piece. In MIG welding, the welder holds the wire feeder, which functions like the alligator clip in arc welding. Instead of using gas flux surrounding the rod, TIG and MIG protect the initial weld from the environment by blowing inert gas onto the weld.

Like arc welding, soldering and brazing use molten metal to join two pieces of metal. However, the metal added during the process has a melting point lower than that of the workpiece, so only the added metal is melted, not the workpiece. Soldering uses metals with a melting point below 800 degrees Fahrenheit; brazing uses metals with a higher melting point. Because soldering and brazing do not melt the work piece, these processes normally do not create the distortions or weaknesses in the work piece that can occur with welding. Soldering commonly is used to join electrical, electronic, and other small metal parts. Brazing produces a stronger joint than does soldering, and often is used to join metals other than steel, such as brass. Brazing can also be used to apply coatings to parts to reduce wear and protect against corrosion.

Skilled welding, soldering, and brazing workers generally plan work from drawings, specifications, or use their knowledge of fluxes and base metals to analyze the parts to be joined. These workers then select and set up welding equipment, execute the planned welds, and examine welds to ensure that they meet standards or specifications. They are even examining the weld while they're welding. By observing problems with the weld, they compensate by adjusting the speed, voltage, amperage, or feed of the rod. Highly skilled welders often are trained to work with a wide variety of materials in addition to steel, such as titanium, aluminum, or plastics.

Some welders have more limited duties, however. They perform routine jobs that already have been planned and laid out and do not require extensive knowledge of welding techniques.

Automated welding is used in an increasing number of production processes. In these instances, a machine or robot performs the welding tasks while monitored by a welding machine operator. Welding, soldering, brazing machine setters, operators, and tenders follow specified layouts, work orders, or blueprints. Operators must load parts correctly and constantly monitor the machine to ensure that it produces the desired bond.

The work of arc, plasma, and oxy-gas cutters is closely related to that of welders. However, instead of joining metals, cutters use the heat from an electric arc, a stream of ionized gas (plasma), or burning gases to cut and trim metal objects to specific dimensions. Cutters also dismantle large objects such as ships, railroad cars, automobiles, buildings, or aircraft. Some operate and monitor cutting machines similar to those used by welding machine operators. Plasma cutting has been increasing in popularity. Unlike other methods, it can cut a wide variety of metals, including stainless steel, aluminum, and titanium.

WORKING CONDITIONS

Welding, soldering, and brazing workers often are exposed to a number of hazards, including the intense light created by the arc, poisonous fumes, and very hot materials. They wear safety shoes, goggles, hoods with protective lenses, and other devices designed to prevent burns and eye injuries and to protect them from falling objects. They normally work in well-ventilated areas to limit their exposure to fumes. Automated welding, soldering, and brazing machine operators are not exposed to as many dangers; however, a face shield or goggles usually provides adequate protection for these workers.

Welders and cutters may work outdoors, often in inclement weather, or indoors, sometimes in a confined area designed to contain sparks and glare. Outdoors, they may work on a scaffold or platform high off the ground. Additionally, they may be required to lift heavy objects and work in a variety of awkward positions, while bending, stooping, or standing to perform work overhead.

Although about 52 percent of welders, solderers, and brazers work a 40-hour week, overtime is common, and some welders work up to 70 hours per week. Welders may also work in shifts as long as 12 hours. Some welders, solderers, brazers, and machine operators work in factories that operate around the clock, necessitating shift work.

TRAINING, OTHER QUALIFICATIONS, AND ADVANCEMENT

Training for welding, soldering, and brazing workers can range from a few weeks of school or on-the-job training for low-skilled positions to several years of combined school and on-the-job training for highly skilled jobs. Formal training is available in high schools, vocational schools, and postsecondary institutions, such as vocational-technical institutes, community colleges, and private welding schools. The armed forces operate welding schools as well. While some employers provide basic training, they prefer to hire workers with experience or more formal training. Courses in blueprint reading, shop mathematics, mechanical drawing, physics, chemistry, and metallurgy are helpful. An understanding of electricity is also very helpful, and knowledge of computers is gaining importance, especially for welding, soldering, and brazing machine operators, who are becoming more responsible for the programming of computer-controlled machines, including robots.

Some welders become certified, a process whereby the employer sends a worker to an institution, such as an independent testing lab, equipment manufacturer, or technical school, to weld a test specimen according to specific codes and standards required by the employer. Testing procedures are based on the standards and codes set by industry associations with which the employer may be affiliated. If the welding inspector at the examining institution determines that the worker has performed according to the employer's guidelines, the inspector will then certify that the welder being tested is able to work with a particular welding procedure.

Welding, soldering, and brazing workers need good eyesight, hand-eye coordination, and manual dexterity. They should be able to concentrate on detailed work for long periods and be able to bend, stoop, and work in awkward positions. In addition, welders increasingly need to be willing to receive training and perform tasks in other production jobs.

Welders can advance to more skilled welding jobs with additional training and experience. For example, they may become welding technicians, supervisors, inspectors, or instructors. Some experienced welders open their own repair shops.

EMPLOYMENT

Welding, soldering, and brazing workers held about 429,000 jobs in 2004. Of these jobs, more than 6 of every 10 were found in manufacturing. Jobs were concentrated in fabricated metal product manufacturing, transportation equipment manufacturing (motor vehicle body and parts, and ship and boat building), machinery manufacturing (agriculture, construction, and mining machinery), architectural and structural metals manufacturing, and construction. Most jobs for welding, soldering, and brazing machine setters, operators, and tenders were found in the same manufacturing industries as skilled welding, soldering, and brazing workers.

JOB OUTLOOK

Employment of welding, soldering, and brazing workers is expected to grow more slowly than average for all occupations over the 2004-2014 period. Despite this outlook, job prospects should be excellent as employers report difficulty finding enough qualified people. In addition, many openings are expected to arise as a large number of workers retire over the next decade.

The major factor affecting employment of welders is the health of the industries in which they work. The manufacturing sector, which employs the most welding, soldering, and brazing workers, is expected to continue to decline as more manufacturing moves overseas. Because almost every manufacturing industry uses welding at some stage of manufacturing or in the repair and maintenance of equipment, this overall decline will affect the demand for welders, although some industries will fare better than others. The construction industry is expected to have solid growth over the next decade and an increasing demand for welders. Government funding for shipbuilding as well as for infrastructure repairs and improvements are expected to generate additional welding jobs.

Pressures to improve productivity and hold down labor costs are leading many companies to invest more in automation, especially computer-controlled and robotically controlled welding machinery. This will reduce the demand for some welders, solderers, and brazers because many repetitive jobs are being automated. The growing use of automation, however, should increase the demand for welding, soldering, brazing machine setters, operators, and tenders. Welders working on construction projects or in equipment repair will not be affected by technology change to the same extent, because their jobs are often unique and not as easily automated.

Despite slower-than-average job growth, technology is creating more uses for welding in the workplace and expanding employment opportunities. For example, new ways are being developed to bond dissimilar materials and nonmetallic materials, such as plastics, composites, and new alloys. Furthermore, laser beam and electron beam welding, new fluxes, and other new technologies and techniques are improving the results of welding, making it useful in a wider assortment of applications. Improvements in technology have also boosted welding productivity, making welding more competitive with other methods of joining materials.

EARNINGS

Median hourly earnings of welders, cutters, solderers, and brazers were $14.72 in May 2004. The middle 50 percent earned between $11.90 and $18.05. The lowest 10 percent had earnings of less than $9.79, while the top 10 percent earned over $22.20. The range of earnings of welders reflects the wide range of skill levels. Median hourly earnings in the industries employing the largest numbers of welders, cutters, solderers, and brazers in May 2004 were:

Motor vehicle parts manufacturing	$16.47
Agriculture, construction and mining machinery mfg.	$14.12
Architectural and structural metals manufacturing	$13.98
Commercial and industrial machinery and repair maintenance	$13.45
Motor vehicle body and trailer manufacturing	$13.45

Median hourly earnings of welding, soldering, and brazing machine setters, operators, and tenders were $14.32 in May 2004. The middle 50 percent earned between $11.73 and $17.78. The lowest 10 percent had earnings of less than $9.63, while the top 10 percent earned over $23.54. Median hourly earnings in motor vehicle parts manufacturing, the industry employing the largest numbers of welding machine operators in May 2004, were $15.43. Many welders belong to unions. Among these are:

- International Association of Machinists and Aerospace Workers

- International Brotherhood of Boilermakers, Iron Ship Builders, Blacksmiths, Forgers and Helpers

- International Union, United Automobile, Aerospace and Agricultural Implement Workers of America

- United Association of Journeymen and Apprentices of the Plumbing, Pipefitting, Sprinkler Fitting Industry of the United States and Canada

- United Electrical, Radio, and Machine Workers of America

RELATED OCCUPATIONS

Welding, soldering, and brazing workers are skilled metal workers. Other metal workers include the following: machinists, machine setters, operators, computer control programmers and operators, tool and die makers, sheet metal workers, and boilermakers. Assemblers and fabricators of electrical and electronic equipment often assemble parts using solder.

SOURCES OF ADDITIONAL INFORMATION

For information on training opportunities and jobs for welding, soldering, and brazing workers, contact local employers, the local office of the state employment service, and schools providing welding, soldering, or brazing training.

Information on careers and educational opportunities in welding is available from:

American Welding Society, 550 N.W. Lejeune Rd., Miami, FL 33126. Internet: http://www.aws.org

---

The material presented in this Introduction is courtesy of Bureau of Labor Statistics, U.S. Department of Labor, *Occupational Outlook Handbook*, 2006-07 Edition, Welding, Soldering, and Brazing Workers.

TECHNIQUES FOR STUDYING AND TEST-TAKING

PREPARING FOR THE EXAM

1. **Make a study schedule.** Assign yourself a period of time each day to devote to preparation for your exam. A regular time is best, but the important thing is daily study.

2. **Study alone.** You will concentrate better when you work by yourself. Keep a list of questions you find puzzling and points you are unsure of to talk over with a friend who is preparing for the same exam. Plan to exchange ideas at a joint review session just before the test.

3. **Eliminate distractions.** Choose a quiet, well-lit spot as far as possible from telephone, television, and family activities. Try to arrange not to be interrupted.

4. **Begin at the beginning.** Read and underline points that you consider significant. Make marginal notes. Flag the pages that you think are especially important with little Post-it™ notes.

5. **Concentrate on the information and instruction chapters.** Study the Code Definitions, the Dictionary of Welding Terms, and the Scrambled Dictionary of Equipment and Usage. Learn the language of the field. Focus on the technique of eliminating wrong answers. This information is important to answering all multiple-choice questions.

6. **Answer the practice questions chapter-by-chapter.** Take note of your weaknesses; use all available textbooks to brush up.

7. **Try the previous exams, if available.** When you believe that you are well prepared, move on to these exams. If possible, answer an entire exam in one sitting. If you must divide your time, divide it into no more than two sessions per exam.

When you do take the practice exams, treat them with respect. Consider each as a dress rehearsal for the real thing. Time yourself accurately, and do not peek at the correct answers. Remember, you are taking these for practice; they will not be scored; they do not count. So learn from them.

IMPORTANT: *Do not memorize questions and answers.* Any question that has been released will not be used again. You may run into questions that are very similar, but you will not be tested with these, exact questions. These questions will give you good practice, but they will not have the exact answers to any of the questions on your exam.

HOW TO TAKE AN EXAM

Get to the examination room about 10 minutes ahead of time. You'll get a better start when you are accustomed to the room. If the room is too cold, too warm, or not well ventilated, call these conditions to the attention of the person in charge.

Make sure that you read the instructions carefully. In many cases, test-takers lose points because they misread some important part of the directions (e.g., reading the incorrect choice instead of the correct choice).

Don't be afraid to guess. The best policy is, of course, to pace yourself so that you can read and consider each question. Sometimes this does not work. Most civil service exam scores, for instance, are based only on the number of questions answered correctly. This means that a wild guess is better than a blank space. There is no penalty for a wrong answer, and you just might guess right. If you see that time is about to run out, mark all the remaining spaces with the same answer. According to the law of averages, some will be right.

Keep in mind that you have bought this book for practice in answering questions. Part of your preparation is learning to pace yourself so that you need not answer randomly at the end. Far better than a wild guess is an educated guess. You make this kind of guess not when you are pressed for time, but when you are not sure of the correct answer. Usually, one or two of the choices are obviously wrong. Eliminate the obviously wrong answers and try to reason among those remaining. Then, if necessary, guess from the smaller field. The odds of choosing a right answer increase if you guess from a field of two instead of from a field of four. When you make an educated guess or a wild guess in the course of the exam, you might want to make a note next to the question number in the test booklet. Then, if there is time, you can go back for a second look.

Reason your way through multiple-choice questions very carefully and methodically.

MULTIPLE-CHOICE TEST-TAKING TIPS

Here are a few examples that we can "walk through" together:

I.   On the job, your supervisor gives you a hurried set of directions. As you start your assigned task, you realize you are not quite clear on the directions given to you. The best action to take would be to:

(a)   Continue with your work, hoping to remember the directions.

(b)   Ask a co-worker in a similar position what he or she would do.

(c)   Ask your supervisor to repeat or clarify certain directions.

(d)   Go on to another assignment.

In this question you are given four possible answers to the problem described. Though the four choices are all possible actions, it is up to you to choose the best course of action in this particular situation.

Choice (a) will likely lead to a poor result; given that you do not recall or understand the directions, you would not be able to perform the assigned task properly. Keep choice (a) in the back of your mind until you have examined the other alternatives. It could be the best of the four choices given.

Choice (b) is also a possible course of action, but is it the best? Consider that the co-worker you consult has not heard the directions. How could he or she know? Perhaps his or her degree of incompetence is greater than yours in this area. Of choices (a) and (b), the better of the two is still choice (a).

Choice (c) is an acceptable course of action. Your supervisor will welcome your questions and will not lose respect for you. At this point, you should hold choice (c) as the best answer and eliminate choice (a).

The course of action in choice (d) is decidedly incorrect because the job at hand would not be completed. Going on to something else does not clear up the problem; it simply postpones your having to make a necessary decision.

After careful consideration of all choices given, choice (c) stands out as the best possible course of action. You should select choice (c) as your answer.

Every question is written about a fact or an accepted concept. The question above indicates the concept that, in general, most supervisory personnel appreciate subordinates questioning directions that may not have been fully understood. This type of clarification precludes subsequent errors on the part of subordinates. On the other hand, many subordinates are reluctant to ask questions for fear that their lack of understanding will detract from their supervisor's evaluation of their abilities.

The supervisor, therefore, has the responsibility of issuing orders and directions in such a way that subordinates will not be discouraged from asking questions. This is the concept on which the sample question was based.

Of course, if you were familiar with this concept, you would have no trouble answering the question. However, if you were not familiar with it, the method outlined here of eliminating incorrect choices and selecting the correct one should prove successful for you.

We have now seen how important it is to identify the concept and the key phrase of the question. Equally, or perhaps even more important, is identifying and analyzing the *key word* or the *qualifying word* in a question—this word is usually an adjective or adverb. Some of the most common key words are:

most	*least*	*best*	*highest*		
lowest	*always*	*never*	*sometimes*		
most likely	*greatest*	*smallest*	*tallest*		
average	*easiest*	*most nearly*	*maximum*		
minimum	*only*	*chiefly*	*mainly*	*but*	*or*

Identifying these key words is usually half the battle in understanding and, consequently, answering all types of exam questions.

Now we will use the elimination method on some additional questions:

II. On the first day you report for work after being appointed as an AC mechanic's helper, you are assigned to routine duties that seem to you to be petty in scope. You should:

   (a) Perform your assignment perfunctorily while conserving your energies for more important work in the future.

   (b) Explain to your superior that you are capable of greater responsibility.

   (c) Consider these duties an opportunity to become thoroughly familiar with the workplace.

   (d) Try to get someone to take care of your assignment until you have become thoroughly acquainted with your new associates.

Once again we are confronted with four possible answers from which we are to select the best one.

Choice (a) will not lead to getting your assigned work done in the best possible manner in the shortest possible time. This would be your responsibility as a newly appointed AC mechanic's helper, and the likelihood of getting to do more important work in the future following the approach stated in this choice is remote. However, since this is only choice (a), we must hold it aside because it may turn out to be the best of the four choices given.

Choice (b) is better than choice (a) because your superior may not be familiar with your capabilities at this point. We therefore should drop choice (a) and retain choice (b) because, once again, it may be the best of the four choices.

The question clearly states that you are newly appointed. Therefore, would it not be wise to perform whatever duties you are assigned in the best possible manner? In this way, you would not only use the opportunity to become acquainted with procedures, but also to demonstrate your abilities.

Choice (c) contains a course of action that will benefit you and the location in which you are working because it will get needed work done. At this point, we drop choice (b) and retain choice (c) because it is by far the better of the two.

The course of action in choice (d) is not likely to get the assignment completed, and it will not enhance your image to your fellow AC mechanic's helpers.

Choice (c), when compared to choice (d), is far better and therefore should be selected as the *best* choice.

Now let us take a question that appeared on a police officer's examination:

III. An off-duty police officer in civilian clothes riding in the rear of a bus notices two teenage boys tampering with the rear emergency door. The most appropriate action for the officer to take is to:

   (a) Tell the boys to discontinue their tampering, pointing out the dangers to life that their actions may create.

   (b) Report the boys' actions to the bus operator and let the bus operator take whatever action is deemed best.

   (c) Signal the bus operator to stop, show the boys the officer's badge, and then order them off the bus.

   (d) Show the boys the officer's badge, order them to stop their actions, and take down their names and addresses.

Before considering the answers to this question, we must accept that it is a well-known fact that a police officer is always on duty to uphold the law even though he or she may be technically off duty.

In choice (a), the course of action taken by the police officer will probably serve to educate the boys and get them to stop their unlawful activity. Since this is only the first choice, we will hold it aside.

In choice (b), we must realize that the authority of the bus operator in this instance is limited. He can ask the boys to stop tampering with the door, but that is all. The police officer can go beyond that point. Therefore, we drop choice (b) and continue to hold choice (a).

Choice (c) as a course of action will not have a lasting effect. What is to stop the boys from boarding the next bus and continuing their unlawful action? We therefore drop choice (c) and continue to hold choice (a).

Choice (d) may have some beneficial effect, but it would not deter the boys from continuing their actions in the future.

When we compare choice (a) with choice (d), we find that choice (a) is the better one overall, and therefore it is the correct answer.

The next question illustrates a type of question that has gained popularity in recent examinations and that requires a two-step evaluation.

First, the reader must evaluate the condition in the question as being "desirable" or "undesirable." Once the determination has been made, we are then left with making a selection from two choices instead of the usual four.

IV. A visitor to an office in a city agency tells one of the office aides that he has an appointment with the supervisor of the office who is expected shortly. The visitor asks for permission to wait in the supervisor's private office, which is unoccupied at the moment. For the office aide to allow the visitor to do so would be:

   (a) Desirable; the visitor would be less likely to disturb the other employees or to be disturbed by them.

   (b) Undesirable; it is not courteous to permit a visitor to be left alone in an office.

   (c) Desirable; the supervisor may wish to speak to the visitor in private.

   (d) Undesirable; the supervisor may have left confidential papers on the desk.

First of all, we must evaluate the course of action on the part of the office aide of permitting the visitor to wait in the supervisor's office as being very undesirable. There is nothing said of the nature of the visit; it may be for a purpose that is not friendly or congenial. There may be papers on the supervisor's desk that he or she does not want the visitor to see or to have knowledge of. Therefore, at this point, we have to decide between choices (b) and (d).

This is definitely not a question of courtesy. Although all visitors should be treated with courtesy, permitting the visitor to wait in the supervisor's office in itself is not the only possible act of courtesy. Another comfortable place could be found for the visitor to wait.

Choice (d) contains the exact reason for evaluating this course of action as being undesirable, and when we compare it with choice (b), choice (d) is far better.

A STRATEGY FOR TEST DAY

On the exam day assigned to you, allow the test itself to be the main attraction of the day. Do not squeeze it in between other activities. Arrive rested, relaxed, and on time.

In fact, plan to arrive a little bit early. Leave plenty of time for traffic tie-ups or other complications that might upset you and interfere with your test performance.

Here is a breakdown of what occurs on examination day and tips on starting off on the right foot and preparing to start your exam:

1. In the test room, the examiner will hand out forms for you to fill out and will give you the instructions that you must follow in taking the examination. Note that you must follow instructions exactly.

2. The examiner will tell you how to fill in the blanks on the forms.

3. Exam time limits and timing signals will be explained.

4. Be sure to ask questions if you do not understand any of the examiner's instructions. You need to be sure that you know exactly what to do.

5. Fill in the grids on the forms carefully and accurately. Filling in the wrong blank may lead to loss of veterans' credits to which you may be entitled or to an incorrect address for your test results.

6. Do not begin the exam until you are told to begin.

7. Stop as soon as the examiner tells you to stop.

8. Do not turn pages until you are told to do so.

9. Do not go back to parts you have already completed.

10. Any infraction of the rules is considered cheating. If you cheat, your test paper will not be scored, and you will not be eligible for appointment.

11. Once the signal has been given and you begin the exam, read every word of every question.

12. Be alert for exclusionary words that might affect your answer-words like "not," "most," and "least."

MARKING YOUR ANSWERS

Read all the choices before you mark your answer. It is statistically true that most errors are made when the last choice is the correct answer. Too many people mark the first answer that seems correct without reading through all the choices to find out which answer is *best*.

Be sure to read the suggestions below now and review them before you take the actual exam. Once you are familiar with the suggestions, you will feel more comfortable with the exam itself and find them all useful when you are marking your answer choices.

1. Mark your answers by completely blackening the answer space of your choice.

2. Mark only ONE answer for each question, even if you think that more than one answer is correct. You must choose only one. If you mark more than one answer, the scoring machine will consider you wrong even if one of your answers is correct.

3. If you change your mind, erase completely. Leave no doubt as to which answer you have chosen.

4. If you do any figuring on the test booklet or on scratch paper, be sure to mark your answer on the answer sheet.

5. Check often to be sure that the question number matches the answer space number and that you have not skipped a space by mistake. If you do skip a space, you must erase all the answers after the skip and answer all the questions again in the right places.

6. Answer every question in order, but do not spend too much time on any one question. If a question seems to be "impossible," do not take it as a personal challenge. Guess and move on. Remember that your task is to answer correctly as many questions as possible. You must apportion your time so as to give yourself a fair chance to read and answer all the questions. If you guess at an answer, mark the question in the test booklet so that you can find it easily if time allows.

7. Guess intelligently if you can. If you do not know the answer to a question, eliminate the answers that you know are wrong and guess from among the remaining choices. If you have no idea whatsoever of the answer to a question, guess anyway. Choose an answer other than the first. The first choice is generally the correct answer less often than the other choices. If your answer is a guess, either an educated guess or a wild one, mark the question in the question booklet so that you can give it a second try if time permits.

8. If you happen to finish before time is up, check to be sure that each question is answered in the right space and that there is only one answer for each question. Return to the difficult questions that you marked in the booklet and try them again. There is no bonus for finishing early so use all your time to perfect your exam paper.

With the combination of techniques for studying and test-taking as well as the self-instructional course and sample examinations in this book, you are given the tools you need to score high on your exam.

WELDING LICENSING EXAM
STUDY GUIDE

Part A

WELDING SAFETY

Welding Licensing Exam Study Guide

—NOTES—

Chapter 1

WELDING SAFETY PRACTICES

Welding Licensing Exam Study Guide

S trict safety measures are necessary during welding and cutting operations to prevent injury to personnel and damage to property. Some of the recommended practices are mandatory and are governed by code requirements. Others are based on shop experience.

There are many variables involved in welding operations. The subject of safety must be treated in a rather general way here. Standards published by the American Welding Society (AWS), American National Standards Institute (ANSI), and other organizations discuss detailed and specific situations.

PERSONNEL PROTECTION

Personnel working near welding stations need to wear proper clothing and use proper equipment to protect themselves from burns, spatter, and radiant energy of the arc. Proper equipment and clothing should be used by welders, helpers, and observers.

Eye and Face Protection

The welding arc should never be observed at close quarters with unprotected eyes. The nearest safe distance for viewing an arc with the unprotected eye is at least 40 feet. Failure to observe this rule can result in various degrees of eye burn. These burns do not usually cause permanent injury, but they can be very painful for several days after exposure.

A helmet-type head shield is standard equipment for protecting the welder's face and eyes from the direct rays of the arc. A hand-held face shield is convenient for use by observers. These shields are generally made from a nonflammable insulating material. They are colored black or gray to minimize reflection and are shaped to protect the face, neck, and ears from the arc. Sunglasses or gas-welding goggles are not adequate protection.

Shields are equipped with a standard-size glass window through which the welder observes the work in the area of the arc. A proper glass lens screens almost 100 percent of infrared and ultraviolet rays and most of the visible rays from the welder's eyes. The lens should be made from tempered glass that is free from bubbles, waves, or other flaws. The flat lens surfaces should be smooth and parallel except for one that is ground for vision correction.

A piece of ordinary colored glass may look like a welding lens, but it would not necessarily have the light-screening characteristics needed for eye protection. Special equipment is required to measure the amount of infrared and ultraviolet rays a lens will absorb. Welding lenses should be purchased only from suppliers who can be depended upon to furnish quality products.

A number of lenses are available in a number of shades for various types of work. Recommended shade numbers for common welding and cutting operations are listed in Table 1.1. Note that the shade number varies with the electrode-size range in shielded-metal-arc welding and with the thickness range of material when in oxygen cutting.

A clear glass that is treated to resist damage from spatter is used to cover the lenses in head and face shields. This protects from breakage and from spatter from the cover lens or plate. It covers the exposed surface of the lens.

In addition to a helmet or shield, goggles with side shields should also be worn during arc welding or cutting operations. Goggles provide protection from spatter and rays from adjacent operations. This is particularly important at times when the shield is removed, as is necessary when replacing electrodes, removing slag, or inspecting the weld. Goggles should be worn by welders' helpers, foremen, inspectors, and others

TABLE 1.1 Recommended lens shade numbers for use in various welding operations.

Welding or cutting operation	Electrode size (in.)	Shade number
Shielded metal-arc welding	1/16–5/32	10
	3/16–1/4	12
	5/16–3/8	14
Gas tungsten-arc welding		
Ferrous	—	12
Nonferrous	—	11
Gas metal-arc welding		
Ferrous	1/16–5/32	12
Nonferrous	1/16–5/32	11
Atomic hydrogen welding	—	10–14
Carbon-arc welding	—	14
Oxygen cutting		
Light–to 1 inch	—	3–4
Medium–1 to 6 inches	—	4–5
Heavy–over 6 inches	—	5–6

Data from AWS A6.2-68 and ANSI Z49.1-67.

working near the arc to protect their eyes from occasional flashes. Goggles should be lightweight, ventilated, and sterilizable, and the frames should be made of a heat-insulating material. Clear, spatter-resistant cover glasses and tinted lenses are used in the goggles.

Submerged-arc welders do not need head shields, but they should use goggles to protect against an accidental flash through the flux.

Protective Clothing

Ultraviolet energy from welding and cutting operations produces a skin burn that, like a sunburn, is not immediately apparent. Thus, welders should wear clothing to protect all exposed skin areas. Woolen clothing is preferred over cotton or synthetic-fiber clothing because it is less easily ignited and offers better protection from rapid changes in temperature. Any cotton or synthetic-fiber clothing worn for welding should be chemically treated to reduce its flammability. Dark-colored shirts are recommended to minimize reflection of rays under the helmet. Outer clothing, such as overalls, should be free from oil or grease. Low shoes with unprotected tops should not be worn. Shoes should be high-topped, steel-toed safety shoes.

Shirt collars and cuffs should be kept buttoned, and pockets should be removed from the front of overalls. These measures prevent sparks or hot metal from lodging in the clothing. For the same reason, legs of trousers or overalls should not be turned up on the outside.

Minimum additional protection commonly used by welders consists of flameproof gauntlet gloves and a flameproof apron made of leather, asbestos, or other suitable material.

For heavy work, fire-resistant leggings or high boots and leather sleeves or a full jacket should be used. For overhead operations, a leather cape or shoulder cover is essential. A leather skull cap should also be worn under the helmet to prevent head burns. Ear protection is desirable for overhead welding or for operations in confined areas. Wire-screen ear protectors are recommended for such operations.

FIRE PREVENTION

Fires connected with welding or cutting operations are usually caused by failure to keep combustible materials away from the work area. Fires rarely occur at permanent production-welding facilities. Most fires involve portable equipment in areas not properly isolated or protected. If the work can be moved, it should be taken to a safe place—such as a fireproof booth for cutting or welding. If the work cannot be moved from a hazardous location, welding should not be done until the area has been made safe. (See Bulletins: NFPA, published by the National Fire Protection Association.)

Welding or cutting should not be done in potentially explosive atmospheres. Atmospheres containing mixtures of flammable gas, vapor, liquid, or dust with air or near stored ignitable materials are considered unsafe. A safe distance for welding in the general area of combustible materials is considered to be 35 feet.

If relocation of combustible materials is not practical, they should be protected with flameproofed covers or shielded with metal or asbestos screens or curtains. Edges of covers at the floor should be sealed to prevent sparks from getting underneath. Portable screens are used for isolating welding and cutting operations and for protection of personnel from rays, spatter, and sparks.

Combustible floors in temporary welding areas should be swept clean, then protected with metal or other noncombustible material. The floor may be protected with damp sand or simply wet with water, but special care must be taken with a wet floor to protect welders and other personnel from being shocked.

All openings in floors or walls should be closed so that any combustible materials on the floor below and in adjacent rooms are not exposed to sparks from the welding operation.

When welding must be done close to wooden construction or other combustible material, additional protection should be provided in the form of suitable fire-extinguishing equipment. This could consist of buckets of water or sand, a connected fire hose, or a portable chemical fire extinguisher, depending on the nature and the amount of combustible material at hand. Workers should be encouraged to be fire-conscious and to be on the lookout for potential fire hazards. It is good practice to have supervisory personnel walk through the welding areas after production operations have been shut down and to have night watchmen give immediate attention to such areas when they come on duty.

PRECAUTIONS IN WELDING CONTAINERS

Tanks, vessels, or other closed containers that have held combustible materials or gases should not be welded or cut unless they have been properly cleaned and marked as safe. Combustible materials include not only the common volatile petroleum products but also:

- Acids that react with metals to produce hydrogen
- Normally nonvolatile oils or solids that can release hazardous vapors when heated
- Fine, dust like particles of a combustible solid that are potentially explosive

Acceptable cleaning methods for such containers include:

- water cleaning
- hot chemical-solution cleaning
- steam cleaning

The method used depends, of course, on the type of material that must be removed from the vessel. Details of approved cleaning procedures are given in AWS standards. Containers that have been cleaned and proved safe should be tagged or stenciled with the words "safe for welding and cutting" and should bear the date and the name of the person who certified the safety.

A supplementary precaution—filling the container with water or inert gas—is recommended for welding or cutting containers even after they have been cleaned by approved methods. The container, of course, must be vented or open so the water or gas can drive out any dangerous fumes. Acceptable inert gases for this purpose are:

- carbon dioxide (CO_2)

- nitrogen

Carbon dioxide is available in pressure cylinders and in solid form as dry ice. Because the gas is heavier than air, it sinks to the bottom of containers that have top openings and, as more gas is added, replaces the air or lighter-than-CO_2 fumes.

Minimum concentration of CO_2 in a container is 50 percent for most work and 80 percent if the previously contained material or vapor was principally hydrogen, carbon monoxide, or acetylene. Thorough flushing with CO_2, however, is desirable practice when the container has held such explosive gases. Recommended minimum concentrations of nitrogen are 60 and 90 percent, respectively, for the same general conditions. Procedures for filling and maintaining minimum concentrations of inert gases in containers for welding and cutting operations are detailed in AWS standards.

VENTILATION

The respiratory-health hazards associated with welding operations are caused principally by:

- inhalation of gases, dusts, and metal fumes
- type and quantity of toxic fumes in a welding area, which depend on the type of welding being done
- the filler and base metals used
- contaminants on the base metal
- solvents in the air
- the amount of air movement or ventilation in the area

Good ventilation is a primary key to avoiding or minimizing respiratory hazards. When welding and cutting mild steels, natural ventilation is usually considered adequate to remove fumes, provided that:

- the room or welding area contains at least 10,000 cubic feet for each welder
- ceiling height is not less than 16 feet
- cross ventilation is not blocked by partitions, equipment, or other structural barriers

Spaces that do not meet these requirements should be equipped with mechanical ventilating equipment that exhausts at least 2000 cubic feet per second (cfm) of air for each welder, except where local exhaust hoods or booths or air-line respirators are used. Welding or cutting operations that involve fluxes or other materials containing fluorine compounds or that involve toxic metals such as zinc, lead, beryllium, cadmium, or mercury require that hose masks, hose masks with blowers, or self-contained breathing equipment be used. Such equipment should meet U.S. Bureau of Mines standards. Some degreasing compounds such as trichlorethylene and perchlorethylene decompose from the heat and from the ultraviolet radiation of an arc.

The products of decomposition are irritating to the eyes and respiratory system. Parts that have been vapor-degreased should not be welded until all degreasing compound and vapors are completely removed.

Because of the chemical breakdown of vapor-degreasing materials under ultraviolet radiation, arc welding should not be done in the vicinity of a vapor-degreasing operation. Carbon-arc welding and gas-tungsten-arc welding should be especially avoided in such areas, because they emit more ultraviolet radiation than other processes.

Exhaust Hoods and Booths

Local exhaust of welding fumes can be provided by adjustable hoods or by fixed enclosures or booths. Individual movable hoods are particularly suitable for bench welding but can be used for any welding or cutting job provided that the hood can be moved so that it is always close to the joint being welded. These hoods are more economical to operate than a general ventilation system, particularly in cold weather, because they require less replacement air to be brought into the room and heated.

Minimum required air velocity at the zone of welding is 100 fpm when the hood is at its furthest position from the joint being welded. For a 3-inch wide-flanged suction opening, this velocity requires an air volume of 150 cfm at 4 to 6 inches from the arc and 600 cfm at 12 inches from the arc.

A ventilated booth is a second type of local exhaust arrangement. A booth is a fixed enclosure that consists of a top and at least two sides that surround the welding operation. Airflow requirements are similar to those for movable hoods—sufficient to maintain a velocity away from the welder of at least 100 fpm.

Welding in Confined Spaces

Welding in small and restricted spaces—such as tanks, boilers, or small compartments of a ship—require particular care to prevent the accumulation of toxic materials and to ensure that personnel have adequate air for breathing. All air that is exhausted must be replaced with clean air.

Where it is impossible to provide such ventilation in confined spaces, welders and helpers should wear hose masks, hose masks with blowers, or self contained breathing equipment of a type that is approved for such use by the U.S. Bureau of Mines. In addition, a worker should be stationed outside each enclosed or confined space during welding to ensure the safety of those working inside.

Protective Clothing

The manufacturers of welding equipment and supplies offer a broad range of protective equipment for welders, such as leather jackets; cape sleeves with detachable bib; waist, bib, or split-leg aprons; and shirt sleeves with snap fasteners at the wrist and adjustable leather straps at the top of the arm. Pants, shirts, and other clothing should be made of a flame-resistant material. The pants should be without cuffs because they can trap sparks and molten pieces of metal. The clothing must be thick enough to minimize or prevent penetration by the dangerous radiation given off by an arc. The arc rays produce very strong visible and invisible rays (both ultraviolet and infrared) that can burn your eyes and skin. Any exposed skin can be burned quickly by these rays. This radiation cannot be seen, but it is present. It results in skin burns similar to sunburn.

Sleeve cuffs should be tight against the wrist to prevent trapping flying sparks or molten particles. For this purpose, elastic bands or gauntlet cuffs are recommended.

Black, flame-resistant cotton twill is often recommended for use with inert-gas-arc welding. Protective clothing made from this type of cloth is cheaper and lighter than leather. Wear high-top leather shoes, work shoes, or boots. Tennis shoes are not acceptable footwear.

MULTIPLE-CHOICE EXAM

1.   Safe work environments are necessary because:

   a. people can get hurt in industrial situations

   b. people enjoy working in rustic looking places

   c. people have no problem with working dangerously

   d. people love to be hurt in order to receive sympathy

2.   Attitudes toward working safely is best learned:

   a. through experience   **b.** trial and error

   c. in the classroom   **d.** in the laboratory

3.   When people learn safety procedures, they develop an (a) _____ concerning safety.

   a. habit   **b.** attitude

   c. mental program   **d.** none of the above

4.   Most accidents are:

   a. by chance   **b.** by plan

   c. caused   **d.** none of the above

5.   In order to work safely, workers should learn:

   a. to organize their work procedures and develop a plan of action

   b. to correctly use tools

   c. to be cooperative

   d. to work hard

6.   A worker is responsible for wearing the proper protective _____.

   a. helmet   **b.** shoes

   c. gloves   **d.** equipment

7.   Safety in welding means knowing the ways to protect yourself from these common injuries and hazards:

   a. burns   **b.** cuts

   c. inadequate ventilation   **d.** none of the above

   e. all of the above

8.   One of the most common injuries that occur in welding is:

    a. eye injuries   **b.** burns

    c. cuts   **d.** pinches

9.   If clothing catches fire when a welder is working the flames should be extinguished by:

    a. running real fast away from the welding area

    b. using water

    c. patted out with a glove

    d. none of the above

10.   In order to protect the eyes, a welder should wear a _____ with a number 10 shade of colored lens.

    a. shoe   **b.** glove

    c. helmet   **d.** hat

11.   Ventilation is a safety concern in a _____ area.

    a. woodworking   **b.** metalworking

    c. welding area   **d.** plastics area

12.   All types of welding give off _____.

    a. sparks   **b.** smoke

    c. gases   **d.** water

13.   A welding booth should have good _____.

    a. light   **b.** sound control

    c. ventilation   **d.** none of the above

14.   During spot welding you should wear:

    a. steel toed shoes   **b.** leather gloves

    c. face shield   **d.** none of the above

    e. all of the above

15.   If arc welders are well maintained, they are not an electrical _____ hazard.

    a. short   **b.** open

    c. shock   **d.** none of the above

16.   Spot welding of galvanized steel will create spatters that can direct molten metal into the welder's
_____.

    a. shoe   **b.** apron pocket

    c. face   **d.** eyes

17.   When working with galvanized steel it is best for safety's sake to wear:

    a. a face shield   **b.** a pair of gloves

    c. steel toed shoes   **d.** all of the above

18.   If arc welders are well maintained they are not an electrical shock _____.

    a. hazard   **b.** unit

    c. cause   **d.** problem

19.   In tungsten inert gas welding, it is possible to get a _____ from the high frequency starting
current.

    a. noise   **b.** problem

    c. shock   **d.** nudge

20.   Small fires which do not create a great amount of heat can be controlled quickly by spraying with:

    a. carbon dioxide   **b.** sand

    c. mud   **d.** water

21.   Carbon dioxide is a good method of fighting these fires:

    a. oil   **b.** gasoline

    c. electrical fires   **d.** all of the above

    e. none of the above

22.   First aid is the immediate _____ given a person who is suffering as the result of an accident.

    a. information   **b.** program

    c. action   **d.** treatment

23. Using a fire _____ is the first action to take to extinguish any fire on a person's hair or body.

    a. hose   **b.** blanket

    c. sponge   **d.** vest

24. First degree burns in which the skin is reddened may be _____ and covered with a burn ointment.

    a. heated   **b.** aerated

    c. treated   **d.** cooled

25. A person in shock may collapse or may merely seem dazed or _____.

    a. stunned   **b.** scalded

    c. hallucinating   **d.** just sick

26. Shock is a very dangerous condition and must be treated _____.

    a. slowly   **b.** with water

    c. with sleep   **d.** quickly

    e. none of the above

TRUE-FALSE EXAM

1. A person in shock may collapse or may merely seem gay and happy.

    True   False

2. Shock is a very serious condition and must be treated immediately.

    True   False

3. First aid is the immediate treatment given to a person who is suffering as the result of an accident.

    True   False

4. When people learn safety procedures, they develop an attitude concerning safety.

    True   False

5. Safety procedures require planned movements.

    True   False

6. When welders are inside or crossing between safety lines, they should feel safe.
    True   False

7. A welder is responsible for wearing the correct protective equipment.
    True   False

8. The arc welder's helmet is designed to protect the eyes and face from falling objects and ultraviolet rays.
    True   False

9. Burns are one of the most common injuries when welding or cutting with a torch.
    True   False

10. If clothing catches on fire when welding, it is best to put out the flames by splashing the welder with water.
    True   False

11. At no time should a person look directly at the light produced by arc welding.
    True   False

12. Most eye injuries can be avoided by wearing safety glasses or a welder's helmet.
    True   False

13. Ventilation is a safety concern in the welding area
    True   False

14. Welding booths have ventilators over them to protect the welder from dirt, dust, carbon dioxide, iron oxides, cadmium oxide, lead oxide, and zinc oxides.
    True   False

15. The intense light produced by arc welding can cause burns to exposed skin.
    True   False

16. When setting up for welding, it is necessary to make sure the welding ground wire is clamped to the work so a welding circuit may be established. The welder should also be grounded.
    True   False

17. Electrical fires should be extinguished with water.
    True   False

18.   There are four classes of fires: A, B, C, and D.

    True   False

19.   Put out a class A fire by using plenty of water.

    True   False

20.   Using a fire blanket is the first action to take when a person's hair or body is on fire.

    True   False

MULTIPLE-CHOICE ANSWERS

1. A	7. E	12. C	17. D	22. D
2. A	8. B	13. C	18. A	23. B
3. B	9. C	14. E	19. C	24. D
4. C	10. C	15. C	20. D	25. A
5. A	11. C	16. C	21. D	26. D
6. D				

TRUE-FALSE ANSWERS

1. F	6. F	11. T	16. T
2. T	7. T	12. T	17. F
3. T	8. T	13. T	18. T
4. T	9. T	14. T	19. T
5. T	10. F	15. T	20. T

—NOTES—

Part B

METALS CUTTING

—NOTES—

Chapter 2
WELDING AND CUTTING

TORCH CUTTING

Oxy-Fuel Gas Cutting

The majority of metal cutting is done by oxy-fuel gas cutting (OFC). However, where the metal is not steel, it is merely melted by means of the flame of the oxy-fuel gas torch and blown away to form a gap, or cut. Where steel is being cut, the process is one of rapid oxidation (burning) of iron at high temperatures, and the chemical equation is:

$$3\,Fe + 2O_2 = Fe_3O_4 + heat$$

This reaction does not occur until steel is at approximately 871 degrees C (1600 degrees F). A gas-flame torch is used to raise the metal to the temperature at which burning will start. After combustion starts, the torch continues to supply the oxygen required for the reaction. In theory, no further heat is required, but in most cases some additional heat must be supplied to compensate for heat losses into the atmosphere and the surrounding metal and thus to assure that the reaction progresses, particularly in the desired direction. Under ideal conditions, the area of burning is well confined to conserve the heat of combustion. Just as in cutting through a steel ingot, no supplementary heating is required. A supply of oxygen through a small pipe keeps the cut progressing. This is known as oxygen lance cutting (LOC), and a temperature of about 1204 degrees C (2200 degrees F) has to be achieved in order for this procedure to be effective.

Acetylene is by far the most common fuel used in oxy-fuel gas cutting; thus the process often is called oxyacetylene cutting (OFC-A). The type of torch commonly used is shown in Figure 2.1. The tip contains a circular array of small holes through which the oxygen-acetylene mixture is supplied for the heating flame. A larger hole in the center supplies a stream of oxygen, controlled by a lever valve. The rapid flow of the cutting oxygen not only produces rapid oxidation but also blows the oxides from the cut. If the torch is adjusted and manipulated properly, a smooth cut results, as shown in the top example of Figure 2.4.

The cutting torch can be manipulated manually. However, in most manufacturing it is moved over the desired path by mechanical means. This procedure yields better accuracy and much smoother cut surfaces. Where duplicate or more complex shapes are required, a pantograph-type machine is widely used. It has a tracer mechanism that follows a template or line drawing. Some models are now equipped with numerical tape or computer controls. All of these types produce desired shapes with remarkable accuracy. Accuracies of ±0.38 mm (0.015 inch) are possible, but accuracies of 0.76 to 1.0 mm (0.030 to 0.040 inch) are more common.

Fuel gases other than acetylene also are used for oxy-fuel gas cutting, the most common being natural gas (OFC-N) and propane (OFC-P). Their use is a matter of economics due to special availability. For certain special work, hydrogen may be used (OFC-H).

In preparing plate edges for welding, two or three simultaneous cuts often are made.

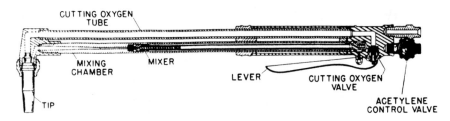

FIGURE 2.1 Oxyacetylene cutting torch. *(Victor)*

Stack Cutting

Precautions must be observed in order to cut a stack of thin sheets of steel successfully; the sheets should be:

- flat and smooth
- free of scale
- clamped together tightly so there are no gaps that could interrupt uniform oxidation and permit slag and molten metal to be entrapped

Stack cutting is a useful technique where a modest quantity of duplicate parts is required, since:

- it is insufficient in number to justify the construction of a blanking die
- the accuracy obtainable is not as good as can be obtained by blanking

Metal Powder Cutting

Not cut readily by the ordinary oxidation process are some ferrous metals: cast iron, stainless steel, and some high alloys. High-melting-point oxides are formed that are not oxidized or melted at the usual temperatures produced in oxyacetylene cutting. They prevent the cutting oxygen from coming into contact with the iron in the material being cut. These difficulties can be eliminated by metal powder cutting (POC). A stream of iron powder is blown into the cutting flame by means of a special opening in the torch tip. The rapid oxidation of this fine iron powder raises the temperature in the cutting area sufficiently to increase the oxidation of the oxides so that cutting can proceed satisfactorily.

Powder cutting also is effective in cutting certain high-melting-point, nonferrous metals, such as copper. That is because these metals do not oxidize readily. To cut most nonferrous materials, it is necessary to melt them and then blow away the molten material. This will form a kerf. These metals frequently have high heat conductivities that make it difficult to maintain a melting temperature. Higher temperatures available with the use of iron powder mean that sufficiently rapid melting of the metal can be achieved so that a satisfactory cut can be made.

Chemical Flux Cutting

Used to cut hard-to cut ferrous metals, a fine stream of a special flux is injected into the cutting oxygen as it leaves the tip of the torch. In this chemical flux cutting (FOC) flux increases the fluidity of the refractory oxides. That is why they are blown from the kerf and leave the iron exposed so that it can be oxidized more readily.

Arc-plasma torches have replaced iron-powder and flux-injection cutting to a considerable extent.

Underwater Torch Cutting

The torch, shown in Figure 2.2, contains an auxiliary skirt that surrounds the main tip. Compressed air flows through the tip. The torch is either ignited in the usual manner before descent or by an electric-spark device after being submerged. Steel can be cut underwater with this special torch. A supply of compressed air provides the secondary oxygen for the oxyacetylene flame. The compressed air also keeps the water away from the zone where the burning of the metal occurs.

Acetylene gas is used for depths up to about 7.6m (25 feet). For greater depths hydrogen is used. That is because the pressure involved is too great for safe operation with acetylene.

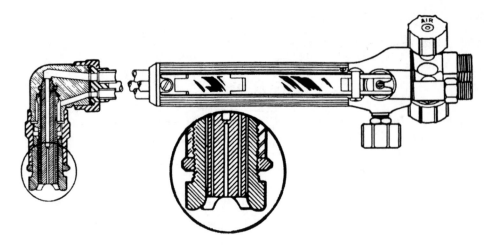

FIGURE 2.2   Underwater torch. *(Bastian-Blessing Company)*

ARC CUTTING

Almost all metals can be cut by the electric-arc process. The metal is melted by the intense heat of the arc and then permitted or forced to flow from the kerf. In one process the resulting edges are considerably rougher than those obtained by oxyacetylene cutting. The rate of cutting is considerably slower. However, some arc-cutting processes can produce edges that are smoother than can be obtained by flame cutting and can do it with much higher cutting speeds. That means that arc cutting processes are very effective and useful for certain applications.

Air-Carbon Arc Cutting

One of the most often used arc cutting processes is air-carbon arc cutting (AAC). An arc is maintained between a carbon electrode and the work piece, and high-velocity jets of air are directed at the molten metal from holes in the electrode holder, as shown in Figure 2.3. While there is some oxidation of molten metal,

FIGURE 2.3   Gun used in arc air process. *(Jackson Products)*

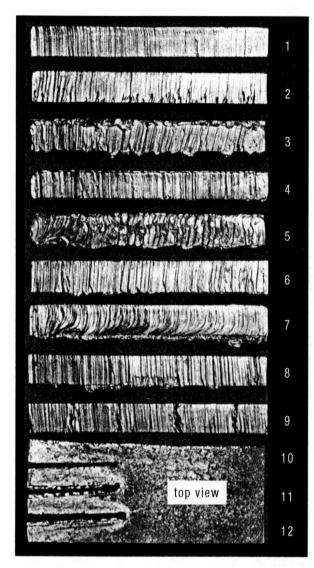

(1) Correct procedure
Compare this cut in 1 inch plate with those below. The edge is square, the drag lines are vertical and not too pronounced.

(2) Preheat flames too small
They are only about 1/8 inch long. Result: cutting speed was too slow, causing bad gouging effect at bottom.

(3) Preheat flames too long
They are about 1/2 inch long. Result: surface has melted over, cut edge is irregular, and there is too much adhering slag.

(4) Oxygen pressure too low
Result: top edge has melted over because of too slow cutting speed.

(5) Oxygen pressure too high
Nozzle size also too small. Result: entire control of the cut has been lost.

(6) Cutting speed too slow
Result: irregularities of drag lines are emphasized.

(7) Cutting speed too high
Result: a pronounced rake to the drag lines and irregularities on the cut edge.

(8) Blowpipe travel unsteady
Result: the cut edge is wavy and irregular.

(9) Lost cut not properly restarted
Result: bad gouges where cut was restarted.

(10) Good kerf
Compare this good kerf (viewed from the top of the plate) with those below.

(11) Too much preheat
Nozzle also is too close to plate. Result: bad melting of the top edges.

(12) Too little preheat
Flames also are too far from the plate. Result: heat spread had opened up kerf at top. Kerf is tapered and too wide.

FIGURE 2.4 Appearance of edges of metal cuts properly and improperly. *(Linde Division)*

the primary effect of the air stream is to blow the molten metal and any oxides away from the cut. The process is used primarily for cutting cast iron and for gouging grooves in steel plates preparatory to welding. Speeds up to 610 mm (24 inches) per minute are readily attained. Although the process can be used for cutting stainless steel and nonferrous metals, plasma arc cutting is more efficient.

The disadvantages of arc air cutting are that it is quite noisy and the hot metal particles tend to be blown out over a substantial area

Plasma Arc Cutting

Torches used in plasma arc cutting (PAC) produce the highest temperature available from any practicable source. They are very useful for cutting metals, particularly nonferrous and stainless types that cannot be cut by the usual rapid oxidation induced by ordinary flame torches. Two types of torches are used, as shown in Figure 2.5. Both are arranged so that the arc column is constricted within a small-diameter nozzle through

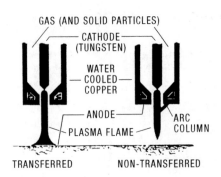

FIGURE 2.5   Principle of arc plasma torches. *(Linde)*

which inert gas is directed. Because the arc fills a substantial part of the nozzle opening, most of the gas must flow through the arc and, as a consequence, is heated to a very high temperature, forming plasma.

With the non-transferred-type torch, wherein the arc column is completed within the nozzle, a temperature of about 16,649 degrees C (30,000 degrees F) is obtainable. With the transferred-type torch, wherein the arc column is between the electrode and the work piece, the temperatures obtainable are estimated to be up to 33,316 degrees C (60,000 degrees F). Obviously, such high temperatures provide a means of very rapid cutting of any material by melting and blowing it from the cut. Cutting speeds up to 7620 mm (300 inches) per minute have been obtained in 6.35 mm (1/4-inch) aluminum. The combination of extremely high temperature and jetlike action of the plasma produces narrow kerfs and remarkably smooth surfaces—nearly as smooth as can be obtained by sawing. Transferred-type torches usually are used for cutting metals, whereas the non-transferred type must be used for nonmetals.

Argon, helium, nitrogen, and mixtures of argon and hydrogen are used. Mixtures of 65 to 80 percent argon and 20 to 35 percent hydrogen are very common. A limited amount of welding is done with plasma torches (PAW) because the temperature is too high for most work. Good results have been achieved with a torch that combines a small non-transferred arc within the torch, which heats the orifice gas, ionizing it and thereby forming a conductive path for the main transferred arc. This permits instant ignition of a low-current arc, which can be lower in magnitude, more stable, and more readily controlled than can an ordinary plasma torch. Separate DC power supplies are used for the pilot and main arcs. An inert shielding gas usually is supplied through an outer cup surrounding the torch.

Gas-Tungsten Arc Cutting

Gas-tungsten arc cutting (GTAC) is used primarily for making holes—up to about 9.5 mm (3/8 inch)—in sheet metal. A special tungsten electrode arc torch is used with a combination of argon and CO_2. Argon gas at low pressure is used while the metal is being melted by the arc; then CO_2, at a pressure of up to 0.56 MPa (80 psi) is used to blow away the molten metal and form the hole.

Metallurgical and Heat Effects from Flame and Arc Cutting

Flame and arc cutting involve high temperatures, often localized. That means it is always possible that the use of these processes could have harmful metallurgical and other heat effects. These possibilities should not be overlooked. In most cases little or no difficulty is experienced, but in others definite steps should be taken to avoid or overcome the harmful effects.

In low-carbon steel (below 0.25 percent carbon) oxyacetylene cutting usually causes no serious metallurgical effects. However, there often is some minor hardening in a thin zone and a small amount of grain growth near the cut. These effects usually will be wiped out if any subsequent welding of the cut edges is done. However, in steels of higher carbon content, these effects can be quite serious. Preheating and/or postheating may be required.

From a heat-effect viewpoint, arc air cutting is about the same as arc welding. If welding follows the use of this process, its heat effects will replace those left by the cutting process, and no special precautions are required. However, if no subsequent welding is to be done, consideration should be given to whether the resulting heat effects will be damaging in view of the load stresses that are anticipated.

Plasma arc cutting is so rapid and the heat is so localized that the heat-affected zone usually is less than 2.38 mm (3/32 inch), and the original hardness of the metal exists beyond 1.59 mm (1/16 inch) from the cut.

All these processes leave some residual stresses, with the cut surface in tension. Except in the case of thin sheets, gas or arc cutting usually will not produce warping. However, if later machining removes only a portion of the total cut surface or not all of the depth in residual tension, the resulting unbalancing of stresses may produce warping. Thus, if later machining is to be done, it may be necessary to remove all cut surfaces to a substantial depth in order to achieve dimensional stability. Machining cuts should be sufficiently deep to get below the hardened surface and avoid dulling tools.

All flame or arc-cut edges are rough to some degree and thus contain geometrical notches that can act as stress raisers and thus reduce the endurance strength. That means if such edges are to be subjected to high or repeated tensile stressing, the cut surface and the heat-affected zone should be machined away or, at a minimum, stress-relief heat treatment must be provided.

LASER-BEAM CUTTING

Any known material can be cut by this process. Laser-beam cutting (LBC) uses intense heat from a laser beam to melt and/or evaporate the material being cut. For some nonmetallic materials the mechanism is purely evaporation. However, for many metals a gas may be supplied to:

- blow away the molten metal and provide a smooth, clean kerf
- speed the process through oxidation by using oxygen (The temperature achieved may be in excess of 11,093 degrees C (20,000 degrees F), and cutting speeds of the order of 25.4 meters (1000 inches) per minute are not uncommon in nonmetals and 508 mm (20 inches) per minute in tough steels.)
- make very accurate cuts (The kerf and heat-affected zone are narrower than with any other thermal cutting process.)
- easily adapt to computer or tape control

WELDING

In oxy-fuel welding, two metals are joined by melting or fusing their adjoining surfaces. This is accomplished by directing an oxy-fuel flame over the metals until a molten puddle is formed. A filler rod may be introduced into the puddle to help the metals form together.

Preparing the Metal to be Welded:

1. Clean the metal joints to be welded of all scale, rust, dirt, paint, and/or grease. Any foreign material that enters the molten puddle will change the metal composition and weaken it.

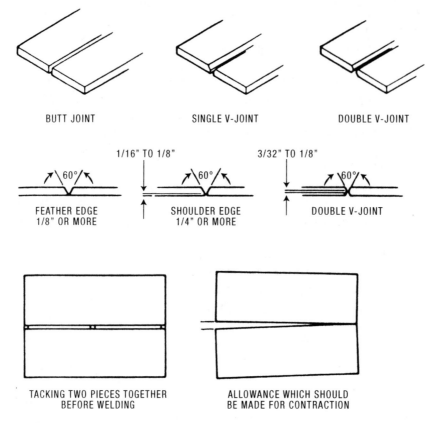

FIGURE 2.6   Preparing the metal. *(Victor)*

2. Base metals 1/8 inch or less do not require beveling, but thicker metals require additional preparation. (See Figure 2.6.)

3. Place the metal to be welded on the work table and determine if and where tacking may be required. Note: as a welding bead is applied, the two pieces of metal may tend to pull together, closing the penetration gap.

To prevent warping:

1. Fuse the ends of the two pieces of metal together before welding. Long pieces may need to be fused every few inches or so along the joint. (See Figure 2.6.)

2. Long pieces may require an additional penetration gap. Add 1/8 to 1/4 inch per linear foot.

Forehand and Backhand Welding Techniques

Two techniques are used for oxy-fuel welding: forehand and backhand. (See Figure 2.7.) The forehand technique is usually recommended for welding material up to 1/8-inch thickness because of better control of the small weld puddle, while backhand welding is generally more suitable for material 1/8 inch and thicker. Increased speed and better fusion at the root of the weld are normally achieved with backhand welding.

In the forehand technique, the welding rod precedes the tip in the direction of the weld. The flame is pointed in the same direction as the weld and is directed downward at an angle, which preheats the edge of the joint. The torch tip and welding rod are manipulated with opposite oscillating motions in semicircular paths to distribute the heat and molten metal uniformly.

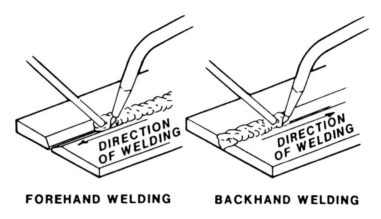

FOREHAND WELDING   **BACKHAND WELDING**

FIGURE 2.7   Forehand and backhand welding. *(Victor)*

In backhand welding, the torch tip precedes the rod in the direction of the weld, and the flame is pointed back toward the molten puddle and completed weld. The end of the welding rod is placed in the flame between the tip and the weld. Less manipulation is required in backhand welding than in forehand welding.

Starting and Finishing the Weld:

The forehand butt weld with filler rod is one of the most common joints made. The basic procedures of the butt weld can be applied to any other type of joint.

1.   Tack or fuse the base metals at the predetermined intervals.

2.   Hold the torch at an angle of approximately 45 degrees to the base metal gap. (See Figure 2.8.)

3.   Move the torch tip over the starting edges of the joint and rotate the flame near the metal in a circular or semicircular motion until the base metals run into a small puddle.

4.   Dip the end of the filler rod in and out of the molten puddle. The molten puddle will melt the rod and add to the puddle.

5.   Continue the dipping motion of the filler rod in the puddle and move the torch back and forth across the penetration gap.

6.   Advance the circular or semicircular motion approximately 1/16 inch per motion until the end of the joint is reached.

7.   Since the angle of the tip flame is preheating the metals ahead of the weld, the last 1/2 inch of weld is critical. Slightly raising the welding tip increase the addition of filler rod to ensure a full, smooth weld.

When You Finish Your Welding Operations:

1.   First shut off the torch oxygen valve, then shut off the torch fuel valve. If this procedure is reversed, a "pop" may occur. The pop throws carbon soot back into the torch and may in time partially clog gas passages.

2.   Close both cylinder valves.

3.   Open the torch-handle oxygen valve and let the oxygen in the system drain out. Close the torch oxygen valve.

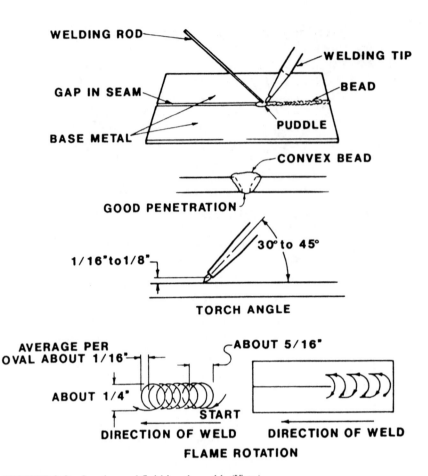

FIGURE 2.8   Starting and finishing the weld. *(Victor)*

4.   Turn the adjusting screw on the oxygen regulator counterclockwise to release all spring pressure.

5.   Open the torch-handle fuel valve and release the pressure in the system. Close the torch fuel valve.

6.   Turn the adjusting screw on the fuel gas regulator counterclockwise to release all spring pressure.

7.   Check the H.P. gauge after a few minutes to be sure the cylinder valve has been turned off completely.

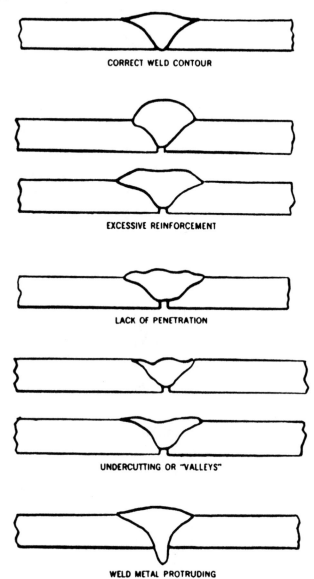

CHARACTERISTICS OF GOOD AND BAD WELDING JOINTS
Compare your weld to the illustrations and determine what improvement could be made.

CORRECT WELD CONTOUR

EXCESSIVE REINFORCEMENT

LACK OF PENETRATION

UNDERCUTTING OR "VALLEYS"

WELD METAL PROTRUDING

FIGURE 2.9   Characteristics of good and bad welding joints. *(Victor)*

MULTIPLE-CHOICE EXAM

1.   Oxidizing means:
     a. too much oxygen

     b. too little oxygen

     c. combining of materials with oxygen

     d. fusing oxygen

2.   Plasma arc cutting means:
     a. melting with a soldering iron

     b. melting metal with a torch

     c. melting with very high temperatures

     d. cutting with a hacksaw

3.   As metals heat and melt, the molecules become free from their neighboring _____ and break the bonds holding them together as a rigid solid.
     a. atoms   **b.** molecules

     c. nucleus   **d.** electrons

4.   Oxygen has a strong _____ affinity for iron.
     a. chemical   **b.** physical

     c. metaphysical   **d.** attractive

5.   What is the critical element in cutting iron or steel?
     a. oxygen   **b.** acetylene

     c. nitrogen   **d.** air

6.   When a narrow area of steel is heated to the temperature of 1400 to 1800 degrees F, a stream of oxygen is directed onto the metal, rapid oxidation takes place, and the oxygen stream _____ the metal along its direction of movement.
     a. cuts   **b.** breaks

     c. welds   **d.** scars

7. The cutting torch and the welding torch are _____.

   a. the same

   b. different in construction

   c. not alike

   d. separate types

8. The cutting torch has _____ needle valves.

   a. one

   b. two

   c. three

   d. four

9. The cutting torch can be used for welding by changing the _____.

   a. handle

   b. tip

   c. hoses

   d. none of the above

10. The cutting operation requires different _____ from those necessary for welding.

    a. hoses

    b. gases

    c. nozzles

    d. pressures

11. The size of the tip and the cutting pressure are related to the _____ of the steel being cut.

    a. thickness

    b. length

    c. height

    d. weight

12. If the oxygen regulator is set for 35 psi with a No. 1-sized cutting tip, what can be cut with it?

    a. 1/2-inch steel plate

    b. 1-inch steel plate

    c. 3/4-inch-inch steel plate

    d. 1-inch cast iron

13. The torch is held at a 45-degree angle to the plate over the edge of the plate until the edge is red hot. Then the torch is moved so that it is _____ degrees from all surfaces.

    a. 180

    b. 60

    c. 30

    d. 90

14. The end of an oxyacetylene cutting tip has a series of holes drilled into it. The center hole is the _____ orifice.

    a. oxygen

    b. acetylene

    c. nitrogen

    d. air

15. When starting and lighting the torch, the oxygen valve on the torch butt is opened:

    a. halfway   **b.** all the way

    c. about a quarter of the way   **d.** none of the above

16. When hand-held cutting tools are used, workers wear a covering to protect their _____.

    a. feet   **b.** feet

    c. hands   **d.** arms

17. In using the torch to cut steel when the edge of the plate has become red-hot, the _____ cutting lever is pressed down to start the cutting.

    a. oxygen   **b.** nitrogen

    c. acetylene   **d.** air

18. The cutting torch is used in cutting work such as removing rivets, gouging metal to prepare for welded joints, and _____ cast iron.

    a. heating   **b.** cutting

    c. molding   **d.** edging

19. The oldest method of shape cutting control is:

    a. mechanical   **b.** automatic

    c. manual   **d.** none of the above

20. Photoelectric tracers are _____ machines that use a photoelectric cell to follow paper drawings of pencil and ink.

    a. burning   **b.** cutting

    c. drawing   **d.** accurate

21. In industrial work, cams and templates are used to _____ contoured cuts for small repetitive work.

    a. shape   **b.** guide

    c. start   **d.** saw

22. Numerically controlled flame-cutting machines are capable of very detailed and _____ cutting.

    a. applied   **b.** free

    c. accurate   **d.** sloppy

23. Plasma arc cutting is used to cut nonferrous metals, _____ steels, refractories, and carbon steels at high speeds.
    a. alloy
    b. stainless
    c. wrought
    d. none of these.

24. Stainless steel up to 2 inches thick can be cut with gas mixtures of argon and _____.
    a. air
    b. oxygen
    c. nitrogen
    d. hydrogen

25. The carbon-dioxide laser is designed to cut, slit, drill, and _____.
    a. draw
    b. weld
    c. circle
    d. slice

26. Laser cutting produces greater _____ over the cutting process.
    a. error correction
    b. quality
    c. control
    d. efficiencies

27. Polarity is defined as:
    a. a pole used in welding
    b. the direction of current flow in a welding circuit
    c. the direction of gas flow in a welding circuit
    d. the label on the probe holding the welding rod

28. Root pass is used to designate:
    a. a weld that needed tacking
    b. a weld on top of the plate being welded
    c. a weld in the bottom of a welded joint
    d. a weld in the top of a welded joint

29. An inert gas is:
    a. a gas that does not react chemically with surrounding metals and materials
    b. a gas that reacts chemically with surrounding metals and materials
    c. a gas that doesn't reach with anything
    d. a gas that reacts with all metals

30. Bonding is the fastening of two pieces of metals by _____.

    a. kinetic energy   **b.** glue

    c. plasma   **d.** diffusion

31. Bonding metal by heat is performed by a number of _____ processes.

    a. welding   **b.** plasma

    c. gluing   **d.** sparking

32. Arc-welding safety is primarily concerned with preventing burns, electrical shock, and _____.

    a. radiation   **b.** cuts and bruises

    c. sunburn   **d.** ultraviolet vision problems

33. In arc welding the _____ is the source of heat for melting the metal.

    a. arc stream   **b.** current

    c. resistance   **d.** welding rod

34. The process of welding is dependent primarily on the condition and _____ of the molten pool.

    a. control   **b.** size

    c. manipulation   **d.** color

35. In arc welding how you use an electrode is important. The angle of the electrode is a factor that aids in the formation of the bead and the _____ of the arc blow.

    a. shape   **b.** control

    c. color   **d.** size

36. In arc welding, the speed at which the electrode is moved controls the size and _____ of the weld head.

    a. length   **b.** contour

    c. color   **d.** shape

37. There are four common positions of welding: horizontal, vertical, overhead, and _____.

    a. flat   **b.** right side

    c. left side   **d.** sideways

38. In shield welding, _____ polarity has higher melting and deposition rates than other types of current.

    a. wrong   **b.** straight

    c. reverse   **d.** none of these

39. In alternating current the polarity _____.

    a. is steady   **b.** never changes

    c. stays the same   **d.** reverses

40. The _____ of the electrode and the position of the weld determine the setting of the welding machine.

    a. size   **b.** shape

    c. composition   **d.** color

41. The welding _____ is one factor that controls the penetration and characteristics of the metal deposit.

    a. current type   **b.** polarity

    c. type   **d.** speed

42. What does a deep-fat-frying sound indicate when welding?

    a. the correct arc length is achieved

    b. the incorrect arc length is achieved

    c. the rod is too big

    d. the rod is too little

43. Thick metals may require a _____ weld.

    a. multipass   **b.** short

    c. stacked   **d.** long

44. Gas metal welding is performed in a _____ atmosphere around a bare wire electrode that melts in the arc.

    a. weak   **b.** controlled

    c. gas   **d.** strong

45.   Welding distortion is controlled by keeping the heat concentration low:

  a. by balancing the heat in the workpiece

  b. by mechanically overpowering the stress forces with clamps and fixtures.

  c. both of the above

  d. neither of the above

46.   The gas-metal-arc welding process has _____ different welding applications.

  a. no   **b.** many

  c. few   **d.** some

47.   Short-circuit transferring of metal takes place when the welding wire electrically shorts, pinches off, arcs to fuse metal to the work, and then _____.

  a. fades out   **b.** reshorts

  c. opens   **d.** spreads

48.   Tubular wire welding has the flux on the _____ of the filler rod tube.

  a. inside   **b.** outside

  c. tip   **d.** clamped end

49.   Flux-cored wire processes with a carbon-dioxide gas shield are operated with a high-arc density or _____ transfer.

  a. spray   **b.** scatter

  c. cool   **d.** temporary

50.   The gas-tungsten arc produces a _____ -quality weld.

  a. high   **b.** low

  c. medium   **d.** none of the above

51.   Reverse polarity produces an oxide film-cleaning action and a broad weld bead with a _____ penetration.

  a. deep   **b.** shallow

  c. medium   **d.** none of these

TRUE-FALSE EXAM

1. Plasma arc-weld surfacing technology helps to produce better-quality products at a competitive market price.

    True   False

2. Electron-beam welding equipment is precision equipment, because the work piece may be moved by 0.001 inch in location and is controlled automatically.

    True   False

3. Electron-beam welding makes a hole in the metal, and as the beam is moved along the work, the liquid metal moves behind the beam and the surface tension of the liquid metal causes the metal from the sides of the weld to intermingle and to be joined upon freezing.

    True   False

4. Electron-beam welding needs a vacuum chamber in order to work properly.

    True   False

5. Laser welding is done by a burst of light that is concentrated into a diameter of 0.001 inch.

    True   False

6. The laser welder can fuse refractory metals, ceramics, and very thick material without warpage or heat damage to adjacent parts.

    True   False

7. Weld nugget refers to the fused area in the metal of a spot weld.

    True   False

8. Hard surfacing means melting a very hard metal powder and welding it in the surface of a base metal.

    True   False

9. Adhesion bonding refers to the fastening of two pieces of metals by diffusion.

    True   False

10. Arc-welding processes generate large amounts of heat, ultraviolet rays, infrared rays, and flying sparks.

    True   False

11.   Arc-welding cables should be checked periodically for wear, cuts, and hot spots.
          True   False

12.   The fumes generated by welding are not harmful to the welder.
          True   False

13.   A chipping hammer is used to make small metal chips for welding.
          True   False

14.   Shielded-metal arc welding is better known as manual arc welding.
          True   False

15.   The process of arc welding is dependent primarily on the condition and manipulation of the molten pool.
          True   False

16.   The angle of the electrodes aids in the formation of the bead and the control of the arc blow.
          True   False

17.   In arc welding the correct speed of travel of the rod will produce a bead that is about one-half the rod diameter high and one and one-half times the rod diameter wide.
          True   False

18.   A fillet weld is made by welding the intersection of two surfaces at 45 degrees or other angles to each other.
          True   False

19.   Welding electrodes are made of steel wires of various compositions with a flux covering the electrode.
          True   False

20.   The American Welding Society has a code for welding-electrode designation.
          True   False

21.   Reverse polarity delivers maximum penetration under standard welding conditions.
          True   False

22.   If you hear a deep-fat-frying sound when arc welding, you know that you are not welding properly.
          True   False

23. Distortion is caused by unequal heat in the work, which in turn produces stress in the weld arc

True   False

24. In order to obtain maximum penetration with a high deposition rate, two different electrodes may be used on heavy welds.

True   False

25. Flux-cored arc welding is a development of the gas-metal arc process.

True   False

MULTIPLE-CHOICE ANSWERS

1. C	12. A	22. C	32. A	42. A
2. C	13. D	23. B	33. A	43. A
3. B	14. A	24. D	34. C	44. B
4. A	15. B	25. B	35. B	45. C
5. A	16. C	26. C	36. B	46. B
6. A	17. A	27. B	37. A	47. B
7. D	18. B	28. C	38. B	48. A
8. C	19. A	29. A	39. D	49. A
9. B	20 A	30. D	40. A	50. A
10. D	21. B	31. A	41. B	51. B
11. A				

TRUE-FALSE ANSWERS

1. T	8. T	14. T	20. T
2. T	9. T	15. T	21. T
3. T	10. T	16. T	22. F
4. T	11. T	17. F	23. T
5. F	12. F	18. F	24. T
6. F	13. F	19. T	25. T
7. T			

Chapter 3

OXYACETYLENE CUTTING

By far the majority of metal cutting is done by oxy-fuel gas cutting (OFC). In a few cases, primarily where the metal is not steel, the metal is merely melted by means of the flame of the oxyfuel gas torch and blown away to form a gap, or cut, in the metal. However, where steel is being cut, the process is one of rapid oxidation (burning) of iron at high temperatures according to the chemical equation:

$$3\ Fe\ +\ 2O_2\ =\ Fe_3O_4\ +\ heat$$

The oxy-fuel cutting process is accomplished by preheating the metal to a bright cherry red, then introducing a stream of cutting oxygen. This will ignite and burn the metal as well as carry away the slag or oxidized residue. Oxy-fuel cutting can be applied to plain carbon steels, low-alloy steels, and other ferrous metals. Non-ferrous metals, stainless steels, and cast iron are not usually cut using oxy-fuel equipment. (See Table 3.1.)

SETTING UP

Here are the procedures for setting up for the cutting process:

1. Inspect the cone end, coupling nut, and torch head for oil, grease, or damaged parts. Inspect the cone end for missing or damaged O-rings.
    Warning: If oil or grease is detected or damage is evident, do not use the apparatus until it has been cleaned and/or repaired by a qualified technician. There must be two O-rings in good condition on the cone end. The absence of either of these O-rings allows premixing of oxygen and fuel gases, which can lead to backflash within the cutting attachment. (See Figure 3.1)

2. Inspect the cutting tip and cutting attachment head. The tapered seating surfaces must be in good condition. If dents, burns, or burned seats are present, resurface the seat. If the cutting attachment is used with poor seating surfaces, backfire or backflash may occur.
    Warning: These seating surfaces prevent premature mixing of gases that can cause fires and explosions. If the tapered seats on the tip are damaged, don't use them. Inspect the preheat and cutting oxygen holes. Splatter can stick on or in these holes. If holes are clogged or obstructed, clean them out with the proper size of tip cleaner.

TABLE 3.1   Common cutting processes and their AWS designations.

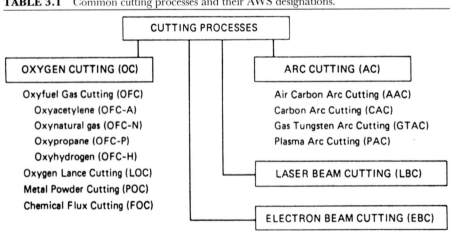

Common cutting processes and their AWS designations.

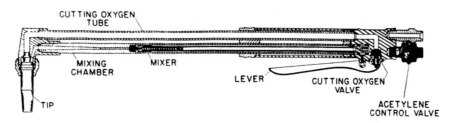

FIGURE 3.1 Oxy-fuel torch. *(Victor)*

3. Connect the cutting attachment to the torch handle and tighten the coupling nut, hand-tight only. Wrench tightening may damage O-rings and create a faulty seal.

4. Refer to welding-nozzle and tip flow data charts for correct cutting tip, regulator pressures, and travel speed.

5. Insert the tip in the cutting-attachment head and tighten the tip nut securely with a wrench (15 to 20 pounds torque).

6. Follow cylinder and regulator safety and operating procedures. Open the oxygen valve on the torch handle completely.

7. Open the preheat oxygen control valve on the cutting attachment and adjust the oxygen regulator to the desired delivery pressure.

8. Close the preheat oxygen control valve.

9. Open the fuel valve on the torch handle and adjust the fuel-regulator delivery range.

10. Close the fuel-control valve on the torch handle.

11. Momentarily depress the cutting oxygen lever to purge the cutting oxygen passage.

12. Open the fuel valve on the torch handle approximately 1/8 turn and ignite the gas with a spark lighter. Be sure the spark lighter is away from the tip and not obstructing the gas flow. **Note:** Wear protective clothing and goggles to shield the eyes from bright light.

13. Continue to increase the fuel supply at the torch handle until the flame stops smoking.

14. Slowly open the preheat oxygen-control valve on the cutting attachment until the preheat flames establish a sharp inner cone. The configuration of the short inner cone is called the neutral flame.

15. Depress the cutting-oxygen lever. If the preheat flame changes slightly to a carburizing flame, continue to depress the cutting-oxygen lever and increase the preheat oxygen at the cutting attachment until the preheat flames are again neutral. If the preheat flames are not the same size and the cutting oxygen is not straight, turn off the torch, let it cool, and clean the tip. **Warning:** If you experience a backfire or backflash (flame disappears and/or emits a shrill hissing sound when it is burning inside the cutting attachment), immediately turn off the preheat oxygen-control valve on the cutting attachment, then turn off the torch-handle fuel valve. Allow the cutting attachment to cool before attempting to relight. If the problem persists, contact your authorized repair station.
Important: Inspect the areas where molten metal and sparks will fall. Serious fires and explosions can be caused by careless torch operations. Take all possible precautions—have fire extinguishers available and remove or protect flammable substances including oxygen and fuel hoses before starting to work.

16. Hold the cutting attachment or torch handle comfortably in both hands so that the back of one hand stabilizes the cutting-tip preheat flames approximately 1/8 inch from the base metal and the other hand is free to depress the cutting-oxygen lever.

17. Direct the preheat flame to the spot where the cut is to be started. Before cutting action can start, the base metal must be preheated to a bright cherry red. When the red spot appears, depress the cutting oxygen lever slowly and fully.

18. When the cut starts, move the torch in the direction you wish to cut.
    Note: Moving too slowly will allow the cut to fuse together, and moving too fast will not allow the metal to be preheated and the cut will be lost.

19. Continue to depress the cutting-oxygen lever past the final edge of the base metal for a good drop cut.

PIERCING

Here are the steps for piercing:

1. Preheat a small spot on the base metal to a bright cherry red.

2. Tilt the torch tip slightly to one side to prevent the sparks and slag from blowing toward you.

3. When the metal is pierced, straighten the torch and move it steadily in the direction you wish to cut. **Note:** If the metal is not pierced all the way through, it probably means that not enough cutting-oxygen pressure has been used.

FINISHING

Here's how to finish your cutting operations:

1. First shut off the torch oxygen valve, then shut off the torch fuel valve. If this procedure is reversed, a "pop" may occur. The "pop" throws carbon soot back into the torch and may in time partially clog gas passages.

2. Close both cylinder valves.

3. Open the torch-handle oxygen valve and release the pressure in the system. Open the cutting oxygen. Close the oxygen valve.

4. Turn the adjusting screw on the oxygen regulator counterclockwise to release all spring pressure.

5. Open the torch-handle fuel valve and release the pressure in the system. Close the fuel valve.

6. Turn the adjusting screw on the fuel-gas regulator counterclockwise to release all spring pressure.

7. Check the H.P. gauge after a few minutes to be sure the cylinder valve is shut off completely.

8. Use a chipping hammer or brush to remove any slag left on the cut edge. Never use the torch head or cutting tip to remove slag from the cut edge.

MAINTAINING CHECK VALVES

Warning: Reverse-flow and internal check valves are mechanical devices that can leak when dirty or damaged. Inspect check valves at least every 6 months—more often if hoses are frequently disconnected. Careless usage, dirt, or abuse can shorten the service life of check valves, thus requiring more frequent inspection. Be sure to test the check valves in a well-ventilated area! The escaping gases create conditions for fires and explosions. Allow no smoking or open flame in the test area.

Regulator Reverse-Flow Check Valves

To test the regulator-type reverse-flow check valve (models CRF & CRO), follow these steps:

1. Turn both regulator adjusting screws counterclockwise so that no pressure will be delivered.

2. Disconnect hoses from the reverse-flow check valves and torch.

3. Remove both reverse-flow check valves from the regulators.

4. Connect the hoses to the regulators.

5. Connect the torch end of the hoses to the reverse-flow check valves.

6. Adjust both regulators until a reading of 2-5 psig is obtained on the low-pressure gauge.

7. Submerse the ends of both reverse-flow check valves in water for 15 seconds or cover the ends with an approved leak-detector solution.

8. Bubbles will appear if a reverse-flow check valve is leaking.

9. If a reverse-flow check valve is leaking, turn off the pressure to the valve. Remove the valve and reinstall in the normal position on the regulator. Flush for 3 to 5 seconds with either 10-psi fuel gas or 30-psi oxygen. depending on the type of reverse-flow check valve used. Remove and retest starting at step 3.

10. If the reverse-flow check valve is still leaking, replace it with a new one.

11. Remove the reverse-flow check valves from hoses and reconnect hoses to the torch. Reinstall the regulator reverse-flow check valve on the regulator outlets and reinstall the hoses on the regulators.

Torch Reverse-Flow and Internal Check Valves

To test internal check valves, follow these steps:

1. Turn both regulator adjusting screws counterclockwise so that no pressure will be delivered.

2. Connect both hoses to the torch.

3. Disconnect one hose from one of the regulators.

4. Open all torch control valves.

5. Plug the end of the tip.

6. Adjust the regulator that is not disconnected until a 2-5 psig reading is obtained on the low-pressure gauge.

7. Put the end of the hose that is disconnected from the regulator under water for 5 seconds or cover it with an approved leak-detector solution.

8. Bubbles will appear if the check valve is leaking. There should be no more than one bubble in 5 seconds.

9. If the check valve is leaking, reconnect the hose to the regulator and unplug the tip. Flush for 3 to 5 seconds with 10-psig fuel gas or 30-psig oxygen, depending on the check valve to be tested.

10. Retest the internal check valve. If it is still leaking, replace it with a new one. Take the torch to a qualified repair technician for check-valve replacement.

11. To test the other internal check valve, reconnect the hose that was disconnected for the first test and disconnect the other hose.

12. Follow the same procedure as for the first test.

13. After both internal check valves have passed the test, purge both the oxygen and fuel lines before lighting the torch. Test hose connections for leaks.[1]

[1] This information courtesy of Victor Equipment Company.

TABLE 3.2 Oxy-fuel cutting reference chart (part 1—description of variables/conditions). (*Victor*)

	Top edge	Bottom edge	Plate face condition and drag line pattern	Sound of cut	Slag pattern	Possible drop cut
Quality cut—all adjustments correct—see Figure 3.2B Part A.	Clean and square with no roll over. No slag on top of plate.	Square and free of slag.	Surface is smooth and clean. Drag lines show uniform, vertical pattern. Plate requires no additional processing.	Smooth and regular.	Regular—consistently vertical through length of cut.	Yes
Travel speed too fast—see Figure 3.2B Part B.	Relatively clean and square.	Considerable slag adheres to bottom edge—cutting oxygen stream moving too fast to allow complete oxidation.	Occasional gouges appear, drag lines are pronounced, and slant away from direction of cut.	No noticeable sputtering.	Irregular pattern as cutting oxygen stream intermittently lags behind the position of tip.	No
Travel speed too slow—see Figure 3.2B Part C.	Rough and uneven. Slightly melted away due to excess pre-heat exposure.	Considerable slag adheres to bottom edge.	Upper portion is clean and smooth. Lower section severely gouged due to wandering oxygen stream.	Erratic—noticeable sputtering.	Irregular and erratic due to uneven progress of oxygen stream.	50/50
Cutting oxygen pressure too high—see Figure 3.2B Part D.	Uneven, out of square. Excessive amount of top edge oxidized as oxygen stream expands upon entry.	Relatively clean and square—free of slag.	Plate is relatively free of pits or gouges but draglines irregular and erratic due to excessive oxygen stream turbulence.	Smooth and regular but exceptionally loud.	Distinct and regular due to force of oxygen stream.	Yes

NOTE: Cutting tip may also be damaged if allowed to contact plate.

(*Continued on next page*)

TABLE 3.2 Oxy-fuel cutting reference chart (part 1–description of variables/conditions), *continued. (Victor)*

	Top edge	Bottom edge	Plate face condition and drag line pattern	Sound of cut	Slag pattern	Possible drop cut
Cutting oxygen pressure too low–see Figure 3.2B Part E.	Generally clean and square.	Considerable slag adheres to plate as cutting oxygen has difficulty penetrating metal.	Plate is fairly smooth but much slag adheres to bottom. Draglines slant away from cut.	Irregular and occasional sputtering.	Irregular and weak–oxygen pressure not sufficient to carry through metal.	No
Too much pre-heat–see Figure 3.2B Part F.	Rounded edge produced by excessive heat. Modern beads also deposited on top of plate.	Moderate amount of slag. Usually adheres to bottom edge.	Draglines are fairly regular and smooth. Excessive metal being removed from plate leaves much slag on bottom edge.	Regular and even but louder than normal (higher flow on pre-heat).	Regular and consistent.	Yes
Too little pre-heat–see Figure 3.2B Part G.	Top edge slightly rounded and out of square.	Often irregular–moderate amount of slag may appear.	Pits and gouges sometimes appear. Draglines uniform and well defined.	Erratic and uneven.	Erratic and uneven.	Not normally
Tip too far from plate–see Figure 3.2B Part H.	Flared and partially blown away–out of square.	Relatively even–little if any slag adhesion.	Smooth and even. Draglines are uniform and vertical.	Smooth and even–constant.	Smooth and even.	Yes
Tip too close to plate–see Figure 3.2B Part I.	Generally rough due to pre-heats interrupting cutting oxygen stream.	Relatively even and slag free.	Occasional gouges will result from pre-heat popping. Draglines show irregular pattern.	Relatively even sputtering.	Usually regular.	50/50

NOTE: Cutting tip may also be damaged if allowed to contact plate.

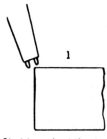

Start to preheat tip at angle on edge of plate.

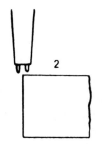

Rotate tip to upright position.

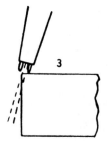

Press oxy valve slowly as cut starts, rotate tip backward slightly.

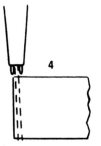

Now rotate to upright position without moving tip forward.

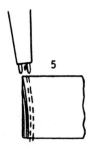

Rotate tip more to point slightly in direction of cut.

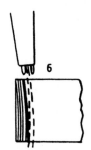

Advance as fast as good cutting action will permit.

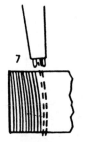

Do not jerk; maintain slight leading angle toward direction of cut.

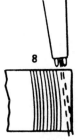

Slow down; let cutting stream sever corner edge at bottom.

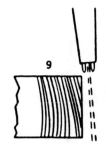

Continue steady forward motion until tip has cleared end.

FIGURE 3.2A Recommended procedures for efficient flame cutting of steel plate. *(Victor)*

(Continued on next page)

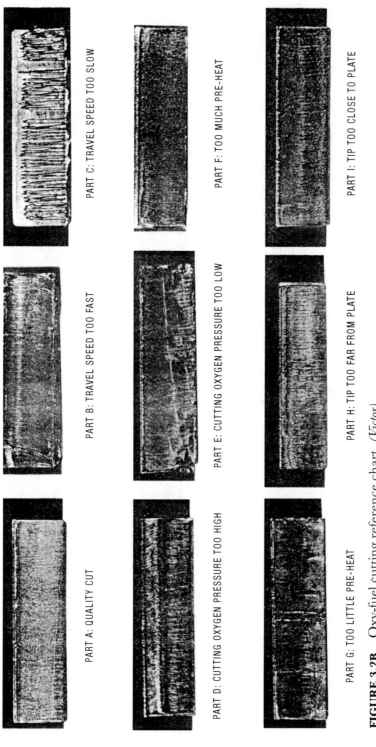

FIGURE 3.2B Oxy-fuel cutting reference chart. (*Victor*)

MULTIPLE-CHOICE EXAM

1. To set up for oxyacetylene welding or cutting, make sure that the tanks of oxygen and acetylene are securely fastened to an immovable _____.

    a. wagon   **b.** cart

    c. truck   **d.** object

2. After the oxygen-tank regulator is attached and cleaned, then attach the regulator on the _____ tank.

    a. acetylene   **b.** propane

    c. nitrogen   **d.** carbon-dioxide

3. After you have attached the regulators to their proper tanks, it is time to attach the _____.

    a. hoses   **b.** torch

    c. blowpipe   **d.** tip

4. The acetylene flame is _____ in color.

    a. blue to white   **b.** red to black

    c. orange to white **d.** yellow to orange

5. The carburizing flame is _____ in color when correctly adjusted.

    a. yellow-orange   **b.** white

    c. orange   **d.** black

6. What are the two parts of the neutral flame?

    a. a white tip and a heat envelope

    b. a blue tip and a reddish heat envelope

    c. both of the above

    d. none of the above

7. When storing hoses for a prolonged time, it is a good idea to cover the ends with:

    a. a piece of tape   **b.** a piece of cork

    c. a piece of asbestos cloth   **d.** a piece of metal

8.   The equipment for cutting metal with the oxyacetylene process is basically the same as for
     _____.

     a. plasma cutting   **b.** electrostatic cutting

     c. welding   **d.** chemical cutting

9.   Each cutting tip will require specific working pressures for both oxygen and acetylene. In all cases the
     working pressure for oxygen will be considerably _____ than the working pressure for acety-
     lene.

     a. lower   **b.** less

     c. more   **d.** none of the above

10. The thickness of the metal being cut, in addition to dictating the cutting tip size, will also determine
     the _____ at which the metal can be cut.

     a. rate   **b.** speed

     c. quality   **d.** none of the above

11. The cutting attachment fits onto the blowpipe _____.

     a. tip   **b.** handle

     c. nipple   **d.** none of the above

12. The oxygen-control knob on the torch handle is opened all the way, but the flow of oxygen is con-
     trolled by the oxygen-control knob on the _____ attachment.

     a. cutting   **b.** welding

     c. tip   **d.** none of the above

13. The proper size of cutting tip is fitted on the _____ attachment.

     a. soldering   **b.** brazing

     c. welding   **d.** cutting

14. The control knob on the torch handle is opened all the way; however, the flow of oxygen is controlled
     by the _____ knob on the cutting attachment.

     a. oxygen   **b.** control

     c. acetylene   **d.** none of the above

15. The flow of acetylene used in cutting operations is controlled by _____ knob(s) on the torch
     handle.

     a. two   **b.** one

     c. three   **d.** none of the above

16. For safety purposes, a bucket of sand and a bucket of _____ should be near or in the cutting work area.

    a. alcohol   **b.** carbon tetrachloride

    c. dirt   **d.** water

17. What is used to mark the metal where it is to be cut?

    a. soapstone pencil   **b.** felt-tip marking pen

    c. scratch awl   **d.** spray can of paint

18. The oxygen cutting lever should _____ be pressed slowly and then held when ready to start the cutting action.

    a. never   **b.** always

    c. sometimes   **d.** none of these

19. You should keep a bucket of water, other than for safety purposes, for:

    a. cooling the cut piece or pieces of metal

    b. aiding in the cutting process

    c. speeding up the cutting process

    d. none of the above

20. A perfectly cut piece of metal looks as if it were cut by a _____.

    a. torch   **b.** hacksaw

    c. hatchet   **d.** hammer

21. If the piece of metal has more heat than is required for cutting, the top of the metal will _____.

    a. indent   **b.** slag up

    c. turn blue   **d.** none of the above

22. When cutting metal, you must depress the oxygen cutting lever just _____ the metal reaches the white-hot stage.

    a. after   **b.** before

    c. when   **d.** none of the above

23. A piece of _____ can be used as a guide when cutting a straight line.

    a. wood   **b.** angle iron

    c. sheet metal   **d.** none of the above

24. What is used to hold the cutting guide in place when cutting a piece of steel?

   a. slip joint pliers   **b.** vise grips

   c. c-clamps   **d.** none of the above

25. When cutting sheet metal, you should use the _____ possible cutting tip.

   a. smallest   **b.** largest

   c. both of the above   **d.** neither of the above

TRUE-FALSE EXAM

1. All cutting attachments are basically the same.

   True   False

2. The six small holes in the cutting torch are called preheat holes.

   True   False

3. The larger center hole is called the cutting hole.

   True   False

4. There are several factors that influence the success of any oxyacetylene cutting process.

   True   False

5. The thickness of the metal being cut, in addition to dictating the cutting-tip size, will also determine the speed at which the metal can be cut.

   True   False

6. The best way to learn how to cut metal with a torch is to watch a single demonstration.

   True   False

7. Each cutting tip will require specific working pressures for both oxygen and acetylene.

   True   False

8. The cutting attachment fits onto the cutting tip.

   True   False

9. Open the valve on the tank of oxygen slowly so the regulator is not damaged by a sudden surge of pressure.

    True   False

10. To start the cutting process, the oxygen control knob is opened all the way.

    True   False

11. The only difference between setting up the oxygen and acetylene lines is that there is only one knob for oxygen.

    True   False

12. Before you begin to cut, you should take a quick look around the work areas to make sure the oxygen and acetylene hoses do not run under the work.

    True   False

13. Oxygen and acetylene hoses should extend from in front of the cutter.

    True   False

14. The reason for keeping the cutting torch 1/8 to 1/4 inch above the metal being cut is to prevent damage to the preheat holes and the oxygen cutting hole.

    True   False

15. Soapstone is simply a type of talc that remains visible under heat.

    True   False

16. Soapstone is used by welders for marking straight and curved cutting lines.

    True   False

17. Use two hands to hold the cutting torch to be able to more accurately cut a straight line.

    True   False

18. One alternative to using a soapstone pencil to mark cuts on metal is to mark the cut line with a series of indentions spaced about 1/4 inch apart.

    True   False

19. Once the edge of the steel plate is cherry red in color, the cutting-oxygen lever is quickly and fully depressed.

    True   False

20. A fast, even motion is best when cutting metal with oxyacetylene equipment.

    True   False

21. There are a number of possibilities as to how the cut will appear: perfect, too slow, too fast, or too much heat are but a few.

    True   False

22. A cut-through steel plate that has been made too quickly will have a buildup of slag like metal on the bottom.

    True   False

23. When the metal has become molten, it will not cut too well unless you first add pure oxygen.

    True   False

24. The time to depress the oxygen-cutting lever is just before the white-hot stage.

    True   False

25. The proper way to add oxygen to cherry-red metal is to very quickly depress the oxygen-cutting lever until the metal is flowing away in an even stream.

    True   False

26. Preheat flames should be from 1/8 to 1/4 inch above the metal being cut.

    True   False

27. A guide, such as angle iron, for an oxyacetylene cutting torch will make it possible for you to make straight cuts on metal.

    True   False

28. The cutting guide can be clamped along the cut line with a pair of vise grips.

    True   False

29. Sheet metal cannot be cut with a cutting torch.

    True   False

30. The thinner the metal, the thinner the cutting tip.

    True   False

31. Steel thicker than 1/4 inch must have beveled edges in order to be oxy-fuel welded.

    True   False

32. The major difference between straight cutting and bevel cutting of steel plate is the angle at which the torch is held.

    True   False

33. You should use a slightly oxidizing flame when cutting a bevel on steel plate.

True   False

34. Cutting thick steel cannot be done with an oxyacetylene cutting torch.

True   False

35. One method used to start a cut in thick metal is to begin at the corner of the metal by slanting the torch in a direction opposite the direction of travel.

True   False

36. When piercing a hole in a piece of steel, there is always a chance of clogging the torch tip.

True   False

37. Beveled edges are not possible when cutting pipe.

True   False

38. Pipe larger than 4 inches is cut with the torch held at a right angle to the cut line.

True   False

39. Cutting small pipe less than 4 inches in diameter requires only a slight torch angle.

True   False

40. A clean cutting tip works best.

True   False

MULTIPLE-CHOICE ANSWERS

1.	D	6.	A	11.	B	16.	D	21.	A
2.	A	7.	A	12.	A	17.	A	22.	B
3.	A	8.	C	13.	D	18.	B	23.	B
4.	D	9.	C	14.	B	19.	A	24.	B
5.	B	10.	B	15.	B	20.	B	25.	A

TRUE-FALSE ANSWERS

1.	T	11.	F	21.	T	31.	T
2.	T	12.	T	22.	T	32.	T
3.	T	13.	F	23.	T	33.	T
4.	T	14.	T	24.	T	34.	F
5.	T	15.	T	25.	F	35.	T
6.	F	16.	T	26.	T	36.	T
7.	T	17.	T	27.	T	37.	F
8.	F	18.	T	28.	T	38.	T
9.	T	19.	F	29.	F	39.	T
10.	T	20.	F	30.	T	40.	T

Chapter 4

PLASMA ARC CUTTING AND WELDING

Torches used in plasma-arc cutting produce some of the highest temperatures available from any practicable source. They are very useful for cutting metals, particularly nonferrous and stainless types. Some types of metals cannot be cut by the usual rapid oxidation induced by ordinary flame torches. Two types of torches are used in plasma cutting. (See Figure 4.1.)

Both of these torches are designed so that the arc column is constricted within a small-diameter nozzle. Through this nozzle passes inert gas. Since the arc fills a substantial part of the nozzle opening, most of the gas flows through the arc and is heated to a very high temperature, thus forming a plasma stream.

In the non-transferred type of torch, the arc column is completed within the nozzle; it has a temperature range of about 16,640 degrees C or 30,000 degrees F.

With the transferred type of torch the arc column is between the electrode and the work. In this type of torch the temperatures are estimated to be up to 33,316 degrees C or 60,000 degrees F. These high temperatures provide a means for very rapid cutting of any material simply by melting and blowing it from the cut. Cutting speeds up to 7620 mm or about 300 inches per minute have been obtained in 6.35-mm (1/4-inch) aluminum. A combination of extremely high temperature and the jetlike action of the plasma produces narrow kerfs and remarkably smooth surfaces. They are nearly as smooth as you can obtain by sawing. Transferred torches usually are used for cutting metals. The non-transferred type is used for nonmetals.

Gases used in this type of cutting are argon, helium, nitrogen, and mixtures of argon and hydrogen. Mixtures of 65 to 80 percent argon and 20 to 35 percent hydrogen are most often used.

A limited amount of welding is done with plasma torches. That is because the temperatures are too high for most work. Good results have been achieved, however, with a torch that integrates a small non-transferred arc within the torch. This heats the orifice gas and ionizes it. The gas forms a conductive path for the main transferred arc. This then permits instant ignition of a low-current arc, which can be lower in magnitude, more stable, and more readily controlled than an ordinary plasma torch. Separate DC power supplies are used for the pilot and main arcs. An inert shielding gas usually is supplied through an outer cup surrounding the torch.

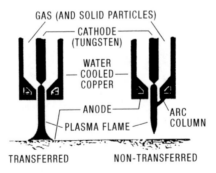

FIGURE 4.1 Principle of arc plasma torches. *(Linde)*

MULTIPLE-CHOICE EXAM

1. An arc is an electric current flowing between two electrodes through an ionized column of gas called a _____.

    a. flame
    b. light source
    c. heat source
    d. plasma

2. The arc column is a mixture of neutral and excited gas _____.

    a. current
    b. atoms
    c. ions
    d. electrons

3. In the central column of the plasma, electrons, atoms, and ions are in accelerated motion and constantly _____.

    a. glowing
    b. moving
    c. colliding
    d. standing still

4. The outer portion of the arc flame is somewhat _____ and consists of recombining gas molecules that were disassociated in the central column.

    a. warmer
    b. hotter
    c. cooler
    d. none of the above

5. In welding, the arc not only provides the heat needed to melt the electrode and the base metal but must under certain conditions also supply the means to transport the molten metal from the tip of the electrode to the _____.

    a. work
    b. arc stream
    c. heat source
    d. none of the above

6. Typical thermal efficiencies for metal-arc welding are in the range of:

    a. 75 to 80 percent
    b. 50 to 60 percent
    c. 10 to 20 percent
    d. 85 to 95 percent

7. Typical thermal efficiencies for welding with non-consumable electrodes are in the range of:

    a. 50 to 60 percent
    b. 75 to 80 percent
    c. 85 to 95 percent
    d. 10 to 20 percent

8. Arc welding may be done with either AC or DC current and with the _____ either positive or negative.

    a. electrode

    b. electrode holder

    c. top piece of work

    d. none of the above

9. A limited amount of welding is done with plasma torches, but the temperature is too _____ for most work.

    a. low

    b. high

    c. expensive

    d. dangerous

10. Good results have been achieved with a plasma torch that integrates a small non-transferred arc within the torch. This heats the orifice gas, ionizing it and thereby forming a conductive path for the main transferred _____.

    a. current

    b. electrode

    c. heat source

    d. arc

TRUE-FALSE EXAM

1. The torches used in plasma-arc cutting produce the highest temperature available from a practicable source.

   True   False

2. Plasma torches are very useful in cutting ferrous materials.

   True   False

3. Plasma torches are very useful in cutting nonferrous and stainless types of metals.

   True   False

4. There are three types of torches for plasma-arc cutting.

   True   False

5. Plasma-arc torches have small-diameter nozzles for directing the gas to the arc.

   True   False

6. One type of plasma cutting torch can obtain temperatures of up to 30,000 degrees F.

   True   False

7. With the transferred type of torch the temperatures can reach as high as 60,000 degrees F.

   True   False

8. Argon is the only gas used with the plasma-arc cutting torch.

   True   False

9. Argon and helium as well as nitrogen are used to operate the plasma-arc torch.

   True   False

10. The transferred type of torch is used for cutting metals.

    True   False

MULTIPLE-CHOICE ANSWERS

1.	D	3.	C	5.	A	7.	A	9.	B
2.	B	4.	C	6.	A	8.	A	10.	D

TRUE-FALSE ANSWERS

1.	T	4.	F	7.	T	9.	T
2.	F	5.	T	8.	F	10.	T
3.	T	6.	T				

Part C

WELDING

Welding Licensing Exam Study Guide

—NOTES—

Chapter 5

SHIELDED-ARC-METAL WELDING

hielded-arc-metal welding is better known as **manual-arc welding.** It is a portable joining process very adaptable to working on large projects such as buildings, bridges, large tanks, and other components or manufactured parts on a site. The work is preformed by a welder striking a 14-inch electrode to the work and melting the work pieces together. The melting of metal from the electrode supplies extra metal to produce the bead.

With this process, an electric arc is struck between the electrically grounded work and a 9- to 18-inch length of covered metal rod—the electrode. The electrode is clamped in an electrode holder, which is joined by a cable to the power source. The welder grips the insulated handle of the electrode holder and maneuvers the tip of the electrode with respect to the weld joint. When he touches the tip of the electrode against the work and then withdraws it to establish the arc, the welding circuit is completed.

The electrode's coating melts. It vaporizes or breaks down chemically nonmetallic substances incorporated in the covering for the shielding. The mixing of molten base metal and filler metal from the electrode provides the coalescence required to effect joining.

As welding progresses, the covered rod becomes shorter and shorter. Finally, the welding must be stopped to remove the stub and replaced with another new electrode. This periodic changing of electrodes is one of the major disadvantages of the process in production welding. It decreases the operating factor, or the percent of the welder's time spent in the actual operation of laying weld beads.

Another disadvantage of shielded-metal-arc welding is the limitation placed on the current that can be used. High amperages, such as those used with semiautomatic guns or automatic welding heads, are impractical because of the long (and varying) length of electrode between the arc and the point of electrical contact between the arc and the electrode holder. The welding current is limited by the resistance heating of the electrode. The electrode temperature must not exceed the "break-down" temperature of the covering. If the temperature is too high, the covering chemicals react with each other or with air and therefore do not function properly at the arc. Coverings with organics will break down at lower temperatures than mineral or low hydrogen types of coverings.

The versatility of the process, however—plus the simplicity of the equipment—are viewed by many users whose work would permit some degree of mechanized welding as overriding its inherent disadvantages. This point of view was formerly well taken, but now that semiautomatic self-shielded flux-cored arc welding has been developed to a simplified (or even superior) degree of versatility and flexibility, there is less justification for adhering to stick-electrode-welding steel fabrication and erection wherever substantial amounts of weld metal must be placed. In fact, the replacement of shielded-metal-arc welding with semiautomatic processes has been a primary means by which steel fabricators and erectors have met price squeezes in their welding operations.

Notwithstanding the limitations of shielded-metal-arc welding, it is certain to remain a primary welding process. It is one well suited in terms of minimal cost of equipment and broad application possibilities for the home mechanic, the farmer, the repair shop, the garage, the trailer-hitch installer, and many others who are concerned entirely with getting a welding job done.

PRINCIPLES OF OPERATION

Take a look at the basic welding circuit and various others created for the shielded-metal-arc process. Note that welding begins when the arc is struck between the work and the tip of the electrode. The heat of the arc melts the electrode and the surface of the work near the arc. Tiny globules of molten metal form on the tip of the electrode and transfer through the arc into the molten weld "pool" or puddle on the work surface.

The transfer through the arc stream is brought about by electrical and magnetic forces. Movement of the arc along the work (or movement of the work under the arc) accomplishes progressive melting and mixing of molten metal, followed by solidification, and thus the unification of parts.

It would be possible to clamp a bare mild-steel electrode into the electrode holder and fuse-join two steel parts. The resulting weld would lack ductility and soundness if judged by present-day standards. The weld metal so deposited would contain oxides and nitrides resulting from reaction of the molten metal with oxygen and nitrogen in the atmosphere.

An essential feature of the electrode used in the shielded-metal-arc process is a covering or coating, applied to the core metal by extrusion or dipping, that contains ingredients to shield the arc and protect the hot metal from chemical reaction with constituents of the atmosphere.

The shielding ingredients have various functions. One is to shield the arc and provides a dense, impenetrable envelope of vapor or gas around the arc and metal to prevent the pickup of oxygen and nitrogen and the chemical formation of oxides and nitrides in the weld puddle. Another is to provide scavengers and deoxidizers to refine the weld metal. A third is to produce a slag coating over molten globules of metal during their transfer through the arc stream and a slag blanket over the molten puddle and the newly solidified weld. Figure 5.1 shows the decomposition of an electrode covering and the manner in which the arc-stream metal is shielded from the air.

Another function of the shield is to provide the ionization needed for AC welding. With alternating current the arc goes out 120 times a second. For it to be reignited each time it goes out, an electrically conductive path must be maintained in the arc stream. Potassium compounds in the electrode covering provide ionized gaseous particles that remain ionized during the fraction of a second that the arc is extinguished with AC hertz reversal. An electrical path for reignition of the arc is thus maintained.

The mechanics of arc shielding vary with the electrode type. Some types of electrodes depend largely on a "disappearing" gaseous shield to protect the arc stream and the weld metal. With these electrodes, only a light covering of slag will be found on the finished weld. Other electrode types depend largely on slag for shielding. The explanation for the protective action is that the tiny globules of metal being transferred in the arc stream are entirely coated with a thin film of molten slag. Presumably, the globules become coated with slag as vaporized slag condenses on them—so the protective action still arises from gasification. In any event, the slag deposits with these types of electrodes are heavy, completely covering the finished weld. Between these extremes are electrodes that depend on various combinations of gas and slag for shielding.

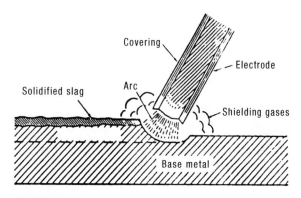

FIGURE 5.1 Gases generated by the decomposition and vaporizatin of materials in the electrode covering, provide a dense shield around the arc stream and over the molten puddle.

The performance characteristics of the electrode are related to their slag-forming properties. Electrodes with heavy slag formation have high deposition rates and are suitable for making large welds downhand. Electrodes that, develop a gaseous shield that disappears into the atmosphere and gives a light slag covering are low-deposition and best suited for making welds in the vertical or overhead positions.

A solid wire core is the main source of filler metal in electrodes for the shielded-metal-arc process. However, the so-called iron-powder electrodes also supply filler metal from iron powder contained in the electrode covering or within a tubular core wire. Iron powder in the covering increases the efficiency of the arc heat and thus the deposition rate. With thickly covered iron-powder electrodes, it is possible to drag the electrode over the joint without the electrode freezing to the work or shorting out. Even though the heavy covering makes contact with the work, the electrical path through the contained powder particles is not adequate in conductivity to short the arc, and any resistance heating that occurs supplements the heat of the arc in melting the electrode. Because heavily covered iron-powder electrodes can be dragged along the joint, less skill is required in their use.

Some electrodes for the shielded-metal-arc process are fabricated with a tubular wire that contains alloying materials in the core. These are used in producing high-alloy deposits. Just as do the conventional electrodes, they have an extruded or dipped covering.

POWER SOURCE

Shielded-metal-arc welding requires relatively low currents (10 to 500 amps) and voltages (17 to 45), depending on the type and size of the electrode used. The current may be either AC. or DC; thus, the power source may be either AC or DC or a combination AC/DC welder. For most work, a variable-voltage power source is preferred, since it is difficult for the welder to hold a constant arc length. With the variable voltage source and the machine set to give a steep volt-ampere curve, the voltage increases or decreases with variations in the arc length to maintain a fairly constant current. The equipment compensates for the inability of the operator to hold an exact arc length, and he or she is able to obtain a uniform deposition rate.

In some welding, however, it may be desirable for the welder to have control over the deposition rate, as when depositing root passes in joints with varying fit-up or in out-of-position work. In these cases variable voltage performance with a flatter voltage-amperage curve is desirable so that the welder can decrease the deposition rate by increasing the arc length or increase it by shortening the arc length. The change from one type of voltage-ampere curve to another is made by changing the open-circuit voltage and current settings of the machine.

The fact that the shielded-metal-arc process can be used with so many electrode types and sizes—in all positions—on a great variety of materials and with flexibility in operator control makes it the most versatile of all welding processes. These advantages are enhanced further by the low cost of equipment. The total advantages of the process, however, must be weighed against the cost per foot of weld when a process is to be selected for a particular job. Shielded-metal-arc welding is a well-recognized way of getting the job done, but too faithful adherence to it often leads to excessive welding costs.

TABLE 5.1 Usability characteristics of mild steel electrodes.

	Type of coating	Position of welding	Type of current* used	Penetration	Rate of deposition	Appearance of bead	Spatter	Stag removal	Minimum tensil strength	Yeild point	Minimum Elongation in 2 inch
E6010	High cellulose sodium	All positions	DC, reverse	Deep	Average rate	Rippled and flat	Moderate	Moderately easy	62,000 psi	50,000 psi	22%
E6011	High cellulose potassium	All positions	AC DC, reverse	Deep	Average rate	Rippled and flat	Moderate	Moderately easy	62,000 psi	50,000 psi	22%
E6012	High titania sodium	All positions	DC, straight AC	Medium	Good rate	Smooth and convex	Slight	Moderately easy	67,000 psi	55,000 psi	17%
E6013	High titania potassium	All positions	AC DC, straight	Mild	Good rate	Smooth and flat to convex	Slight	Easy	67,000 psi	55,000 psi	17%
E7014	Iron powder titania	All positions	AC DC, straight	Mild	High rate	Smooth and flat to convex	Very slight	Very easy	72,000 psi	60,000 psi	17%
E7016	Low hydrogen potassium	All positions	AC DC, reverse	Mild to medium	Good rate	Smooth and flat to convex	Slight	Moderately easy	72,000 psi	60,000 psi	22%
E6020	High iron oxide	Flat hor. fillets	Flat—DC, AC hor. fillets—DC str., AC	Deep	High rate	Smooth and flat to concave	Slight	Easy	62,000 psi	55,000 psi	25%
E7024	Iron powder titania	Flat hor. fillets	DC AC	Deep	Very high rate	Smooth and flat to convex	Very slight	Very easy	72,000 psi	60,000 psi	17%
E6027	Iron powder iron oxide	Flat hor. fillets	Flat—DC, AC hor. fillets—DC str., AC	Deep	Very high rate	Flat to concave	Slight	Very easy	62,000 psi	50,000 psi	25%
E7018	Iron powder low hydrogen	All positions	AC DC, reverse	Mild	High rate	Smooth and slightly convex	Very slight	Very easy	67,000 psi	55,000 psi	22%
E7028	Iron powder low hydrogen	Flat hor. fillets	AC DC, reverse	Mild	Very high rate	Smooth and slightly convex	Very easy	Very easy	67,000 psi	55,000 psi	22%

* DC reverse means DC, reverse polarity (electrode positive).
  DC straight means DC, straight polarity (electrode negative).

MULTIPLE-CHOICE EXAM

1.   Shielded-metal-arc welding is better known as _____- arc welding.

      a. automatic   **b.** submerged

      c. manual   **d.** gas

2.   The melting of the electrode supplies the extra metal to produce the _____.

      a. flash   **b.** bead

      c. arc   **d.** flux

3.   The process of welding is dependent primarily on the condition and _____ of the molten pool.

      a. manipulation   **b.** temperature

      c. size   **d.** color

4.   Four factors being controlled by the welder affect the molten pool: length of arc, setting of the power
      source, angle of the electrode, and _____.

      a. speed of travel   **b.** temperature of the pool

      c. size of the pool   **d.** angle of the electrode

5.   The welding process produces the sound of _____.

      a. frying eggs   **b.** birds chirping

      c. clicking   **d.** whistling

6.   The setting of the power source determines the available _____ for producing the heat.

      a. voltage surge   **b.** resistance

      c. wattage   **d.** amperage

7.   The angle of the electrode is a factor that aids in the formation of the bead and the control of the arc
      _____.

      a. throw   **b.** blow

      c. intensity   **d.** brightness

8.   The correct angle of the electrode is obtained when it is held vertical to the plate lying on the weld-
      ing bench and inclined toward the direction of travel from 0 to _____ degrees.

      a. 10   **b.** 15

      c. 20   **d.** 25

9.   The correct speed of travel will produce a bead that is about one-half the rod diameter high and one and one-half times the rod _____ in width.

  a. height   **b.** length

  c. diameter   **d.** radius

10.   The four welding positions are flat, vertical, overhead, and _____.

  a. forward   **b.** backward

  c. round   **d.** horizontal

11.   The two basic types of joints are fillet and _____.

  a. butt   **b.** groove

  c. double v   **d.** lap

12.   In the vertical welding position the welded seam is perpendicular to the earth's _____.

  a. surface   **b.** moon

  c. magnetic north pole   **d.** magnetic south pole

13.   A welder working around a large pipe lying on the ground will have to weld all around the seam, thus all four welding _____ are used.

  a. electrodes   **b.** motions

  c. types   **d.** positions

14.   Which of the following is NOT a function of electrode flux coatings?

  a. makes the arc easy to start

  b. protects the liquid metal from the oxides and nitrides of the atmosphere

  c. stabilizes the arc for heat control

  d. increases splatter during welding

15.   Welding rods with heavy coatings of iron powder supply additional amounts of _____ metal to the weld.

  a. slag   **b.** filler

  c. fluxed   **d.** oxygen

16. Welding rods or electrodes have numbering systems that have been developed by the American Society for Testing Materials (ASTM) and the _____.

    a. National Association of Manufacturers

    b. American Welding Society

    c. American Society of Welding

    d. Welding Society of America

17. Shielded-metal-arc electrodes start with the letter _____.

    a. A   **b.** E

    c. C   **d.** D

18. Welding polarity refers to the flow of _____.

    a. current   **b.** resistance

    c. gas   **d.** ions

19. In shielded-arc welding, straight polarity has higher melting and deposition rates than other types of _____.

    a. welding   **b.** resistance

    c. voltage   **d.** current

20. Reverse polarity delivers maximum penetration under standard welding conditions and gives an advantage for root passes in grooved welding or where fit-up of the parts demands high heat for _____.

    a. fluxing   **b.** melting

    c. fusion   **d.** pooling

21. In alternating current the polarity is constantly _____.

    a. reversing   **b.** failing

    c. breaking down   **d.** none of the above

22. The size of the electrode and the position of the _____ determine the setting of the welding machine.

    a. welder   **b.** rod

    c. metal   **d.** weld

23. Thick metals may require a _____ weld and thus are joined with a series of welds.

    a. special   **b.** short

    c. long   **d.** multipass

24. To get maximum penetration with a high _____ rate, two different electrodes may be used on heavy metals.

    a. deposition   **b.** speed

    c. electrode   **d.** none of the above

25. Distortion is caused by _____ heat in the work, which in turn produces stress in the weld area.

    a. equal   **b.** unequal

    c. excess   **d.** less

TRUE-FALSE EXAM

1. The shielded-metal-arc process is commonly known as manual or stick-electrode welding.

True   False

2. The shielded-metal-arc welding process is used by home workshops as well as industry.

True   False

3. With this process (shielded-metal-arc) an electric arc is struck between the electrically grounded work and a 9- to 18-inch electrode.

True   False

4. The electrode in arc welding does not melt.

True   False

5. A disadvantage of shielded-metal-arc welding is the limitation placed on the current that can be used.

True   False

6. High amperages, such as those used with semiautomatic guns or automatic welding heads, are practical because of the long and varying length of electrode between the arc and the point of electrical contact in the jaws of the electrode holder.

True   False

7. The welding current is limited by the resistance heating of the electrode.

True   False

8. The electrode temperature must not exceed the breakdown temperature of its covering.

True   False

9. Welding begins when the arc is struck between the work and the tip of the electrode.

True   False

10. The heat of the arc melts the electrode and the surface of the work near the arc.

True   False

11. An essential feature of the electrode used in the shielded-metal-arc welding process is the electrode covering or coating.

True   False

12. A function of the shield is to provide the ionization needed for DC welding.

True   False

13. Potassium compounds in the electrode covering provide ionized gaseous particles that remain ionized during the fraction of a second that the arc is extinguished with AC cycle reversal.

True   False

14. The transfer through the arc stream is brought about by electrical and magnetic forces.

True   False

15. Some types of electrodes depend largely on a disappearing gaseous shield to protect the arc stream and the weld metal.

True   False

MULTIPLE-CHOICE ANSWERS

1.	C	6.	D	11.	B	16.	B	21.	A
2.	B	7.	B	12.	A	17.	B	22.	D
3.	A	8.	D	13.	D	18.	A	23.	D
4.	A	9.	C	14.	D	19.	D	24.	A
5.	A	10.	D	15.	B	20.	C	25.	B

TRUE-FALSE ANSWERS

1.	T	5.	T	9.	T	13.	T
2.	T	6.	F	10.	T	14.	T
3.	T	7.	T	11.	T	15.	T
4.	F	8.	T	12.	F		

—NOTES—

Chapter 6
OXYACETYLENE WELDING

Gas welding has been used commercially throughout the twentieth century, and the twenty-first is still making good use of it. Acetylene gas is easily produced from inexpensive material. It produces an intense heat quickly. It was first used to power the lights on coal miners' helmets.

GAS-WELDING APPLICATIONS

The first welding process used commercially was oxyacetylene. It is good for many applications, in particular:

- Cast iron
- Wrought iron
- Low-alloy steels
- Copper
- Bronze

Except for repair and maintenance work, oxyacetylene welding has been replaced by various arc-welding processes, including shielded-metal-arc welding (or stick welding), gas-metal-arc welding (or **MIG** welding), and tungsten-metal-arc welding (or **TIG** welding). Although relegated to a minor role in welding, oxyacetylene is still used for a wide variety of other non-welding uses, such as:

- Cutting
- Preheating
- Postheating
- Flame hardening
- Case hardening
- Braze welding
- Brazing
- Soldering
- Descaling

OAW advantages include the following:

- Self-contained and easily portable equipment
- Widely available equipment
- Relatively inexpensive equipment
- Easy to learn

OAW disadvantages include:

- Slower welding process than other methods
- Uses volatile and potentially dangerous gases

- Fuel-gas and oxygen cylinders require special handling to avoid damage
- Damaged cylinders can result in fire or explosions

EQUIPMENT

A typical oxyacetylene welding station will include the following components:

- Welding torch and nozzle
- Oxygen cylinder, oxygen regulator, and oxygen hose
- Acetylene cylinder, acetylene supply, acetylene regulator, and acetylene hose
- Flashback arrestors and check valves and torch lighter/sparklighter (Figure 6.1)

Welding Torch

The welding torch is designed to mix oxygen and acetylene in nearly equal amounts and then to ignite and burn the gas mixture at the torch tip. The welding torch has two tubes, one for oxygen and the other for acetylene; a mixing chamber; and oxygen and acetylene valves to control and adjust the flame. (See Figure 6.2.)

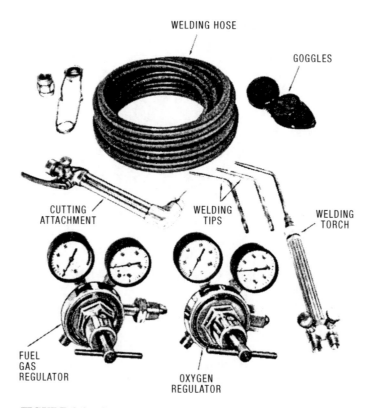

FIGURE 6.1 Components of an oxyacetylene welding station.

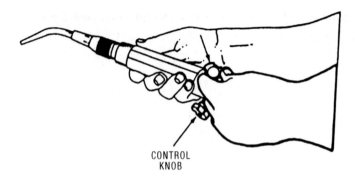

CONTROL
KNOB

FIGURE 6.2   Typical oxyacetylene torch.

Welding-Torch Tips

Welding tips may be purchased in a wide variety of sizes and shapes. The suitability of a particular welding-tip design depends on a number of factors including the accessibility of the area being welded, the rate of welding speed desired, and the size of the welding flame required for the job.

Manufacturers have their own numbering systems to indicate different welding-tip sizes. There is no industry standard, although comparison charts are available. Keep in mind that tip-size identifications have no bearing on minimum or maximum gas consumption or on flame characteristics.

Drill size alone also fails to give an adequate comparison between the various makes of welding tips with identical tip-drill size. The internal torch and tip construction may vary. The gas exit velocities as well as gas-pressure adjustments may also vary.

The welding-torch nozzle is replaceable and is available in a wide variety of sizes. The size selected will depend on the thickness of the metal being welded. Table 6.1 contains data for the selection of welding tips.

TABLE 6.1   Oxyacetylene welding tips data.

Tip size	Drill size	Oxygen pressure psi		Acetylene pressure psi		Acetylene consumption CFH*		Metal thickness
		Min.	Max.	Min.	Max.	Min.	Max.	
000	75	1/4	2	1/2	2	1/2	3	up to 1/32″
00	70	1	2	1	2	1	4	1/64″–3/64″
0	65	1	3	1	3	2	6	1/32″–5/64″
1	60	1	4	1	4	4	8	3/64″–3/32″
2	56	2	5	2	5	7	13	1/16″–1/8″
3	53	3	7	3	7	8	36	1/8″–3/16″
4	49	4	10	4	10	10	41	3/16″–1/4″
5	43	5	12	5	15	15	59	1/4″–1/2″
6	36	6	14	6	15	55	127	1/2″–3/4″
7	30	7	16	7	15	78	152	3/4″–1 1/4″
8	29	9	19	8	15	81	160	1 1/4″–2″
9	28	10	20	9	15	90	166	2″–2 1/2″
10	27	11	22	10	15	100	169	2 1/2″–3″
11	26	13	24	11	15	106	175	3″–3 1/2″
12	25	14	28	12	15	111	211	3 1/2″–4″

* Oxygen consumption is 1.1 times the acetylene under neutral flame conditions. Gas consumption data is merely for rough estimating purposes. It will vary greatly on the material being welded and the particular skill of the operator. Pressures are approximate for hose length up to 25 feet. Increase for longer hose lengths about 1 psi per 25 feet.

These are recommended sizes, and all variables should be taken into consideration before making the final selection. **Note:** Because tips are subject to wear, they must be replaced from time to time. Use an appropriate wrench for this purpose (never pliers). Such malfunctions as "backfire," appropriate tip at the recommended pressure.

Oxygen Cylinders

Oxygen cylinders are seamless steel containers holding about 244 cubic feet of oxygen at a pressure of 2200 psi at 70 degrees F. A smaller cylinder holding about 122 cubic feet is also available. A typical oxygen cylinder (Figure 6.3) is approximately 9 inches in outside diameter, 54 inches in height, and between 104 and 139 pounds in weight (empty). The difference in weight depends upon the type of steel used to construct the cylinder. A full cylinder will increase the weight by about 20 pounds.

The oxygen supplied through these cylinders is about 99.5 percent pure. The oxygen pressure will vary according to temperature changes, but the weight and percentage of oxygen remain the same.

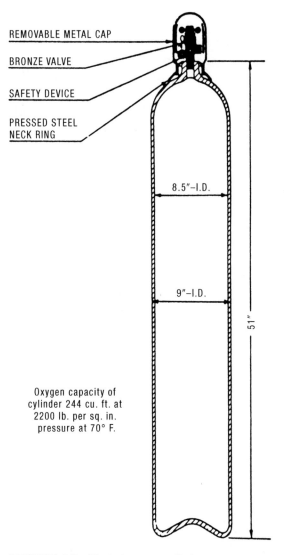

FIGURE 6.3 Typical oxygen cylinder.

TABLE 6.2 Variations in oxygen cylinder pressures with temperature changes.

Temperature degrees F	Pressure psi approx.	Temperature degrees F	Pressure psi approx.
120	2500	30	1960
100	2380	20	1900
80	2260	10	1840
70	2200	0	1780
60	2140	−10	1720
50	2080	−20	1660
40	2020		

Gauge pressures indicated for varying temperature conditions on a full cylinder initially charged to 2200 psi at 70°F. Values identical for 244 cu. ft. and 122 cu. ft. cylinder.

A special bronze valve is inserted into the top of the oxygen cylinder. A removable cap is placed over it for protection. The oxygen-cylinder valve is designed to withstand great pressures. If the pressure becomes too great, a disc in the safety plug located on the oxygen valve will rupture, allowing the excess oxygen to escape before the high pressure ruptures the cylinder. Pressure regulators can be attached to the oxygen-cylinder valve at an outlet with a standard male thread.

Variations in oxygen-cylinder pressures with respect to temperature changes are given in Table 6.2. Oxygen-cylinder content as indicated by gauge pressure at 70 degrees F (for a 244-foot^3 cylinder) is listed in Table 6.3.

Caution: Because oil or grease in contact with oxygen may cause a violent and explosive reaction, resulting in serious injuries or even death, the following precautions must be taken:

- Post a warning note on or near the oxygen cylinder forbidding the use of oil or grease on the cylinder fittings.
- Always keep oxygen cylinders and apparatus free from oil, grease, and other flammable or explosive substances.
- Never handle oxygen cylinders or equipment with oily hands or gloves.

TABLE 6.3 Oxygen cylinder content.*

Gauge pressure psi	Content cu. ft.	Gauge pressure psi	Content cu. ft.
190	20	1285	140
285	30	1375	150
380	40	1465	160
475	50	1550	170
565	60	1640	180
655	70	1730	190
745	80	1820	200
840	90	1910	210
930	100	2000	220
1020	110	2090	230
1110	120	2200	244
1200	130		

* 122-cu. ft. cylinder content one-half above volumes.

Acetylene Cylinders

Acetylene cylinders are available in sizes ranging from a capacity of 10 cubic feet up to 360 cubic feet. Cylinders are available on lease or sold on an exchange basis. Figure 6.4 illustrates the construction of a typical acetylene cylinder. Note the removable cap cover over the valve at the top.

Acetylene can be used safely if it is dissolved under a pressure of 250 psi in liquid acetone. Liquid acetone is used to stabilize the acetylene. Always secure an acetylene cylinder in an upright position on a hand truck or against a wall to prevent acetone leakage.

Never drop or strike an acetylene cylinder. Even a slight shock can cause a cylinder to explode. Acetylene is very unstable at elevated pressures.

Note: A damaged cylinder must be removed immediately to the outside of the building. No chances should be taken with a leaking acetylene-gas cylinder.

Caution: Never exceed a 15-psi regulator gauge reading when reducing the acetylene to a working pressure. The discharge rate of an acetylene cylinder should not be excessive. An excessive rate of discharge results in the drawing off of acetone, which results in strong weakening effects on the weld. The discharge rate should be controlled as a ratio to the total cubic-foot capacity of the cylinder (e.g., a 50-cubic-feet-per-hour discharge rate for a cylinder with a 275-cubic-foot capacity).

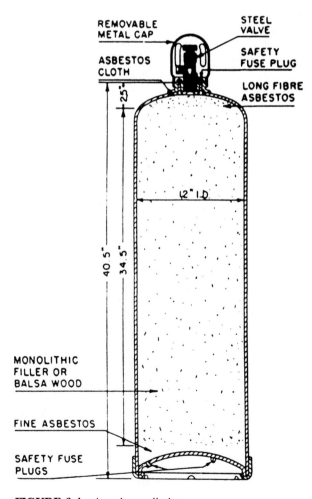

FIGURE 6.4 Acetylene cylinder.

Hoses and Hose Connections

Hoses used in oxyacetylene and other gas-welding processes should be strong, nonporous, flexible, and not subject to kinking. The best hoses are constructed from nonblooming neoprene tubing reinforced with braided rayon. The outer coating should be resistant to oil and grease and tough enough to survive most shop conditions. These hoses may be purchased as single hoses (with 1/8- to 1/2-inch inside diameters) or as twin double-barreled hoses (one line for the oxygen, the other for the fuel gas) with metal binders at the base. Gas hoses can be purchased in continuous lengths up to 300 feet. Standard reel lots are approximately 100 feet. In addition, cut lengths (with connections) are available in boxed lengths of 12 1/2, 25, and 100 feet.

Hose Inspection and Maintenance:

- Regularly inspect the hoses and connections for damage or wear.
- Replace leaky, worn, or damaged hoses. Do not attempt to repair them. A tape repaired hose is not a safe seal.
- Replace damaged hose connections.
- Always blow out the hoses before welding or cutting.
- Keep as much hose off the floor as possible to protect it from being run over by equipment or stepped on.
- Keep hose runs as short as possible to avoid damage.
- Coil excess hose to avoid kinking and tangles in the line.

An oxygen hose is usually black or green (blue in the United Kingdom). The fuel-gas hose is generally red. The color distinctions are standard for safety reasons. For example, using an oxygen hose to carry acetylene could cause a serious accident. No hose should be used for any purpose other than what it was designed for. A further safety precaution is found in the design of the threading on the hose connections. The fuel-gas hoses have left-handed threaded connections (and a groove around the outside). The oxygen hoses are fitted with right-handed thread connections, with swivel nuts at both ends.

Note: Use the shortest possible hose length between the cylinders and the torch. You will use less oxygen and acetylene with shorter hoses than longer ones. You will also experience less pressure drop at the torch when using shorter hoses.

A metal clamp is used to attach the welding hose to a nipple. A nut on the other end of the nipple is connected to the regulator or torch. Sometimes the shape of the nipple will indicate its use. A bullet-shaped nipple is generally used for oxygen hoses and a nipple with a straight taper for gas-fuel hoses. Another identification method is a groove running around the center of the acetylene nut. This is an indication of a left-hand thread, which will only screw into the acetylene outlet.

Note: Multiple sets of hoses may be connected to a single regulator on a single set of oxyacetylene cylinders only by installing an approved commercially available fitting listed by a nationally recognized testing laboratory. This fitting must be installed on the output side of the regulator and must have an integral shutoff and reverse-flow check valve on each branch.

Pressure Regulators

Oxygen and acetylene pressure regulators are used to control gas pressure. They do this by reducing the high pressure of the gases stored in the cylinders to a working pressure delivered to the torch and by maintaining a constant gas working pressure during the welding process. There is a regulator for the oxygen and another for the fuel gas. These gas-pressure regulators are connected between the gas cylinder and the hose leading to the torch.

Regulators may differ according to their capacity (ranging from light to high) and according to the type of gas for which they are designed. For example, an oxygen regulator cannot be used for acetylene gas and vice versa.

Oxygen regulators do not have the same size or thread as acetylene regulators. This is a safety feature that prevents them from being attached to the wrong cylinder. Another safety feature is found in the hose outlet from the regulator. The outlet connections for oxygen hoses have right-hand threads. For acetylene and other fuel gases the outlet connections have left-hand threads.

Both single- and two-stage regulators are available for use in welding systems. Many regulators are constructed with two gauges. One gauge indicates the pressure of the gas in the cylinder (the high-pressure gauge), and the other indicates the working pressure of the gas being delivered to the torch (the low-pressure gauge).

A single-stage regulator requires torch adjustments to maintain a constant working pressure. As the cylinder pressure decreases, the regulator pressure likewise drops, necessitating torch adjustment. In the two-stage regulator, there is automatic compensation for any drop in cylinder pressure. The two-stage regulator is virtually two regulators in one, which operate together to reduce the pressure progressively in two stages instead of one.

The first stage in the two-stage regulator serves as a high-pressure reduction chamber. A predetermined pressure is set and maintained by the spring and diaphragm. The gas (at a reduced pressure) then flows into the second of the two stages, which serves as a low-pressure reduction chamber. In this second chamber, pressure control is controlled by an adjustment screw.

Proper regulator maintenance is important for safe and efficient operation. The following points are particularly important to remember:

- The adjusting screw on the regulator must always be released before opening the cylinder valve. Failure to do this results in extreme pressure against the gauge that measures the line pressure and may cause damage to the regulator.

- Gauge-equipped regulators should never be dropped, improperly stored, or in any way subjected to careless handling. The gauges are extremely sensitive instruments and can be easily rendered inoperable.

- Never oil a regulator. Most regulators will have the instruction "use no oil" printed on the face of both gauges.

- All regulator connections should be tight and free from leaks.

Caution: Never adjust an acetylene regulator to allow a discharge greater than 15-psi (103.4-kPa) gauge.

Flashback Arrestors and Check Valves

Flashback arrestors, also called flame traps, and nonreturn spring-loaded check valves should be installed between the acetylene and oxygen openings in the torch and the matching hoses. **Note:** Some torches are designed with integral flashback arrestors and check valves.

Flashback is a potentially dangerous condition caused by the burning of an oxygen-and-fuel-gas mixture in the mixing chamber of the torch handle instead of at the torch tip. If the burning fuel-gas-oxygen mixture passes through the hoses and regulators to the cylinders, it can result in a fire or an explosion leading to serious injury or even death. Among the causes of flashback are:

- Opening the fuel-gas and oxygen cylinder valves and then attempting to light a torch with a blocked tip

- Loose hose connections and/or hose leaks

- Low gas velocity produced by incorrect gas pressure

- Lighting the torch with a failed oxygen or acetylene regulator

Hose check valves prevent the oxygen and fuel gas from crossing over and mixing together in a volatile mixture at the back of the torch mixing chamber. Flame arrestors, commonly narrow stainless-steel tubes, stop the flame by absorbing its heat and constricting its passage.

Lighters

Oxygen-and-fuel mixtures should not be ignited with a match. A sudden flare-up on ignition could cause the hands or other parts of the body to be burned. To ensure against such danger, the gas mixture should be ignited with a device that provides the required degree of safety for the welder. One such device is a spark-lighter, a simple, inexpensive device made of flint and steel. Some sparklighters are equipped with pistol grips and shoot a shower of sparks at the gas flowing from the torch. Others have a rotating flint holder that permits longer use before having to insert a new flint.

Welding Rods

Gaswelding rods, or filler rods, are thin strips of metal used to add metal to the weld during the welding process. During welding, the filler rod melts and deposits its metal into the puddle, where it joins with the molten base metal to form a strong weld. Because the composition of the filler rod must be matched as closely as possible to that of the base metal, the selection of the appropriate rod for the job is extremely important. Choosing the wrong filler rod will result in a weak and ineffective weld.

Welding rods are available in a variety of sizes and compositions. (See Table 6.4.) The sizes range in diameter from 1/16 to 3/8 inch. The cast-iron rods are sold in 24-inch lengths. All others are available in 36-inch lengths.

TABLE 6.4   Gas welding filler rods.

Type	Comments
RG45	• Copper-coated low carbon steel gas welding rod. • AWS A5.2 Class RG45. • Use a neutral flame to avoid excess oxidation or carbon pick-up. • Commonly used to weld ordinary low carbon steel up to 1/4″ thick. • Recommended where ductility and machinability are most important. Produces high quality welds which are ductile and free of porosity. • Excellent for steel sheet, plates, pipes, castings and structural shapes. • No flux required.
RG60	• Low alloy steel gas welding rod. • AWS A5.2 Class RG60. • Use a neutral flame to avoid excess oxidation or carbon pick-up. • Used to produce high tensile strength quality welds on low carbon and low alloy steels such as sheets, plates, pipes of grade A and B analysis and structural shapes. • Recommended for critical welds that must respond to the same annealing and heat treatment as regular grades of cast steel. • The high silicon and manganese composition removes impurities from the molten metal thereby eliminating the need for flux. • RG60 rod is also used as a filler metal in gas tungsten arc welding (GTAW/TIG).
Bare brass rod	• Low-fuming bronze gas welding rods made of copper tin alloy to produce a rod that flows easily, joins a variety of metals including cast and malleable iron, galvanized steel, brass copper, and steel. • Produce strong joints up to 63, 000 psi tensile strength. • Flux required for bare brass rod.
Flux-coated brass or low-fuming bronze rod	• Same characteristics as bare brass rod. • Flux contained in rod coating.

Flux

Flux is a material used to prevent, dissolve, or facilitate removal of oxides and other undesirable substances that can contaminate the weld. The flux material is fusible and nonmetallic. As a result of the chemical reaction between the flux and the oxide, a slag is formed, which floats to the top of the molten puddle of metal during the welding process. The slag can then be removed from the surface after the weld has cooled.

The following should be kept in mind when dealing with fluxes:

- The chemical composition of a flux depends on the metal or metals on which it is to be applied.
- Fluxes may be divided into three main categories:
  (1) welding fluxes
  (2) brazing fluxes
  (3) soldering fluxes
- Welding fluxes are divided into gas-welding fluxes and braze-welding fluxes.
- Fluxes are sold as powders, pastes, or liquids (frequently in plastic squeeze bottles). Quantities are available from as small as 1/2- to 5-pound cans, in jars to 25 pounds, or in larger sizes sold in drums.
- Some powdered fluxes may be applied by dipping the heated welding rod into the can. The flux will stick to the rod. The flux powder may also be applied directly to the surface of the base metal. Some powdered fluxes are mixed with alcohol or water and applied to the surface as a paste.
- Paste or liquid fluxes are applied in the form in which they are purchased.

Welding-Tip Cleaners

Welding-torch tips must be cleaned regularly in order to prolong the life of the tip and to provide consistently high performance. Stainless-steel tip cleaners (wires) are available in various diameters to fit different tip orifices. All deposits on the inside of the tip must be removed without enlarging the size of the opening (orifice).

EQUIPMENT SETUP

Basic Assembly

Follow these directions to set up your oxyacetylene-welding equipment:

1. Fasten the oxygen and acetylene cylinders in an upright position to a welding cart, wall, or fixed vertical surface to keep them from falling over. Use a chain or a nonflammable material to fasten them in place. Locate them as close as possible to the welding job but away from open flames.

2. Remove the caps from both cylinders. Examine the cylinder outlet nozzles for stripped threads or a damaged connection seat.

3. Open and close the oxygen cylinder valve very quickly to blow out any loose dust or dirt that may have accumulated in the outlet nozzle. Wipe the connection seat with a dry, clean cloth. If the dust or dirt is not removed, it could damage the regulator. These un-removed contaminants can also cause incorrect gauge readings.

Caution: Turn your head away when opening the oxygen cylinder valve and make sure that the oxygen stream is not directed toward another worker, a spark, or an open flame. The high pressure of the oxygen stream can cause serious injury to the eyes. Oxygen will also cause a serious fire if ignited by a spark or flame.

Note: The valve opening and the inlet nipple should be shiny and clean inside and outside. This is particularly important for the oxygen cylinder. Oil or grease in the presence of oxygen is flammable or even explosive. Never allow oxygen to contact oil, grease, or other flammable substances.

4.   Repeat Step 3 for the acetylene cylinder.

5.   Connect the oxygen regulator to the oxygen cylinder.

Note: Regulators must be used only with the gas and pressure range for which they are intended and marked. Cylinder-valve outlets and the matching inlet connections on regulators have been designed to minimize, as far as possible, the chances of making incorrect connections.

6.   Connect the acetylene regulator to the acetylene cylinder.

7.   Connect the green or black oxygen hose to the oxygen regulator. The oxygen hose has a right-hand thread and must be turned clockwise to tighten. Make the connection tight, but avoid overtightening it.

8.   Connect the red acetylene hose to the acetylene regulator. The acetylene hose has a left-hand thread and must be turned counterclockwise to tighten. Again, do not overtighten the connection.

9.   Charge the oxygen regulator by slowly opening the oxygen cylinder valve. Opening the valve slowly prevents damage to the regulator seat.

Caution: Never face the regulator when opening the cylinder valve. A defective regulator may allow the gas to blow through with enough force to break the gauge glass, resulting in possible injury to anyone standing nearby. Always stand to one side of the regulator when opening the cylinder valve and turn the valve slowly.

10.   Open the oxygen-regulator-adjusting screw (the T-handle on the regulator), blow out any dirt or debris in the oxygen hose, and then close it.

11.   Repeat Steps 9 and 10 for the acetylene regulator and hose.

12.   Connect the oxygen hose to the oxygen needle valve on the torch and the acetylene hose to the acetylene needle valve. Again, the oxygen hose has a right-hand thread, whereas the acetylene hose has a left-hand one.

13.   Close the welding-torch-needle valves and open the oxygen and acetylene cylinder valves. Adjust the regulators for a normal working pressure and check for leaks at all the connections with soap and water. Soap bubbles will indicate a leak. Tighten the connections with a wrench. If this fails to eliminate the problem, shut off the oxygen and acetylene, check for stripped threads, defective hoses (old hoses become porous), or other damaged parts, and repair or replace as necessary.

Note: Leaks must be repaired before attempting to use the welding equipment. A leak will not only produce wasted gas, but it can also cause a fire or explosion. Always check the equipment for leaks on a periodic basis, not just when the equipment has been initially set up.

Tip Selection

The tip size will depend on the thickness of the metal being welded. Use a tip with a small opening for welding thin sheet metal, and follow these directions:

Note: It is very important to use the correct tip size, with the proper working pressure. If too small a tip is employed, the heat will not be sufficient to fuse the metal to the proper depth. When the tip is too large, the heat is too great, thereby burning holes in the metal.

1. Select the correct tip size for the welding job. The size of the welding tip depends on many factors, including the thickness of the metal, the welding position, and the type of metal being welded. If the tip is too small for the work, too much time is wasted in making the weld and poor fusion is likely to result. If the tip is too large, it is likely to produce poor metal in the weld (due to overoxidation) and a rough-looking job due to lack of control of the flowing metal.

2. Install the tip in the nozzle.

Torch Lighting

Here's how to light the torch:

1. Point the torch tip down and away from your body.

2. Open the oxygen and acetylene cylinder valves and set the working pressure to correspond to the size of tip being used.

Caution: Always face away from the regulator when opening a cylinder valve. A defect in the regulator may cause the gas to blow through, shattering the glass and blowing it into your face. Remember, oxygen and acetylene are charged in the tanks under a high pressure, and if the gas is permitted to come against the regulator suddenly, it may cause some damage to the equipment.

3. Open the acetylene cylinder valve approximately one complete turn and the oxygen all the way. Next, turn the oxygen and acetylene regulator-adjusting valves to the required working pressures.

4. Open the acetylene needle valve on the torch about one-quarter turn and spark the lighter at the torch tip.

Caution: If you take too long to spark the lighter, acetylene will build up around the torch tip. When the excess acetylene is finally lit, it may cause an explosion, resulting in burns to the hand.

Note: Never use a match to light a torch. This procedure brings your fingers too close to the tip, and the sudden ignition of the acetylene can cause serious burns.

Caution: Make no attempt to relight a torch from the hot metal when welding in an enclosed box, tank, drum, or other small cavity. There may be just enough unburned gas in this confined space to cause an explosion as the acetylene from the tip comes in contact with the hot metal. Instead, move the torch to the open, relight it in the usual manner, and make the necessary adjustments before resuming the weld.

5. After igniting the acetylene gas, make adjustments at the acetylene valve to permit enough gas to burn to give the proper intensity of flame. If the acetylene flame is accompanied by a lot of smoke, increase the amount of acetylene until the smoke disappears and the flame seems to "jump" off the torch tip.

TABLE 6.5 Basic components of the oxyacetylene flame.

Flame components (Figure 6.5)	Comments
Inner tip	Also called the cone. That portion of the flame located at the bore of the torch nozzle. The innermost portion of the flame. Caution: never allow the inner tip of the flame to touch the work because it will burn through the metal.
Beard or brush	Located between the outer envelope and the inner tip or cone. Not always present. Caution: never allow the beard or brush to touch the work.
Outer envelope	Extends around the beard and inner tip. Much larger in volume than the beard and inner tip because it is fed oxygen from the surrounding atmosphere.

6.   When the acetylene flame is properly adjusted, slowly open the oxygen valve to allow the air in the line to escape gradually so that it will not blow out the flame.

7.   The torch oxygen valve is then gradually opened until the flame changes from a ragged yellow flame to a perfectly formed bluish cone. This flame is known as a neutral flame and is the torch flame commonly used for most welding.

Adjusting the Flame

The proportions of oxygen and acetylene can be adjusted to produce a neutral, oxidizing, or carburizing flame. Oxyacetylene welding is normally performed using a neutral flame, produced by mixing roughly equal amounts of oxygen and acetylene. (See Tables 6.5 and 6.6 and Figure 6.5.)

Turning off the Torch

Here's how to turn the torch off:

1.   Close the acetylene-needle valve on the torch first. The acetylene-needle valve is closed first, since shutting off the flow of this gas will immediately extinguish the flame, whereas if the oxygen is shut off first, the acetylene will continue to burn, throwing off smoke and soot.

2.   Close the oxygen-needle valve on the torch.

TABLE 6.6 Oxyacetylene flame types.

Flame type (Figure 6.6)	Comments
Acetylene flame	Very white, large, smoky flame produced when torch is first lit.
Carbonizing flame	Also called a reducing flame. Produced by burning an excess of acetylene. The flame very often has no beard or brush on its inner tip. When adjusted with a small beard, it may be used on most nonferrous metals (those not containing iron elements). The outer envelope is usually the portion of the flame used on these metals.
Neutral or flame	Produced by burning one part acetylene gas and slightly more than one part oxygen.
Oxidizing flame	Produced by burning an excess of oxygen. The flame has no beard and both the inner tip and envelope are shorter. The oxidizing flame is of limited use because it is harmful to many metals.

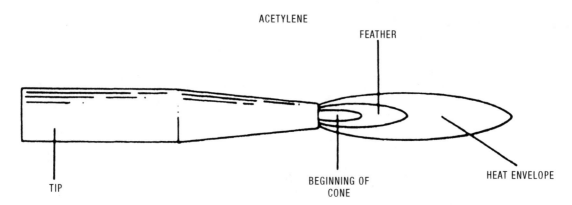

FIGURE 6.5 Three basic components of the oxyacetylene flame.

Equipment Closure

Follow these important steps to shut down your equipment:

1. If the entire welding unit is to be shut down, shut off both the acetylene and the oxygen cylinder valves.

2. Remove the pressure on the working gauges by opening the needle valves until the lines are drained. Then promptly close the needle valves.

3. Release the adjusting screws on the pressure regulators by turning them to the left.

4. Disconnect the hoses from the torch.

5. Disconnect the hoses from the regulators.

6. Remove the regulators from the cylinders and replace the protective caps on the cylinders.

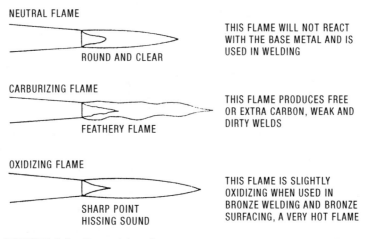

FIGURE 6.6 Oxyacetylene flame types.

WELDING METHODS

The two basic methods of running a weld bead are the forehand welding method and the backhand welding method. They differ according to whether the torch tip is pointed in the direction of the weld bead or back towards the welded seam. In addition to the forehand and backhand welding methods, the welder is also confronted with the problem of the welding position.

Forehand Method

In the forehand welding method, the tip of the welding torch follows the welding rod in the direction in which the weld is being made (Figure 6.7). This method is characterized by wide semicircular movements of both the welding tip and the welding rod, which are manipulated so as to produce opposite oscillating movements. The flame is pointed in the direction of the weld but slightly downward so as to preheat the edges of the joint.

The major difficulty with the forehand welding method is encountered when welding thicker metals. In order to obtain adequate penetration and proper fusion of the groove surfaces and to permit the movements of the tip and rod, a wide V-groove (90-degree included angle) must be created at the joint. This results in a large puddle, which can prove difficult to control, particularly in the overhead position.

Note: Forehand welding is also sometimes referred to as ripple welding or puddle welding.

Backhand Method

In the backhand welding method, the tip of the torch precedes the welding rod in the direction in which the weld is being made (Figure 6.8). In contrast to the forehand welding method, the flame is pointed back at the puddle and the welding rod. In addition, the torch is moved steadily down the groove without any oscillating movements. The welding rod, on the other hand, may be moved in circles (within the puddle) or semicircles (back and forth around the puddle).

Backhand welding results in the formation of smaller puddles. A narrower V-groove (30-degree bevel or 60-degree included angle) is required than is the case with the forehand welding method. As a result, greater control is provided as well as reduced welding costs.

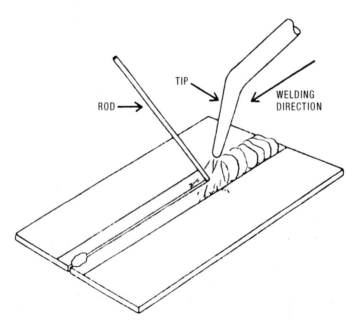

FIGURE 6.7   Forehand welding method. Note: the welding rod moves ahead of the torch tip.

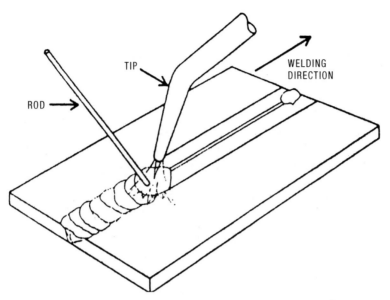

FIGURE 6.8   Backhand welding method. Note: the torch moves ahead of the welding rod.

Filler Metal

The end of the welding rod should be melted by keeping it beneath the surface of the molten weld puddle. Never allow it to come into contact with the inner cone of the torch flame. Do not hold the welding rod above the puddle so that the filler metal drips into the puddle.

Torch Angle and Movement

Figures 6.9 and 6.10 illustrate common torch movements, recommended torch angle and distance of the inner cone tip of the flame from the surface, and the weld bead produced by increasing and decreasing the welding speed.

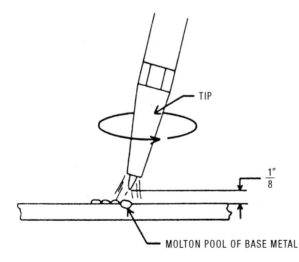

MOLTON POOL OF BASE METAL

The tip of the flame is held about 1/8" above the surface of the metal and is moved in a circle.

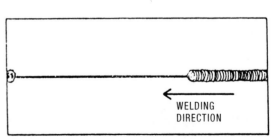

The tip is moved in a crescent motion, carrying the molten pool of metal across the work to be welded. Filler rod is added in the forward edge of the molten pool to add reinforcement to the weld.

FIGURE 6.9   A recommended torch angle and inner cone distance from the metal surface.

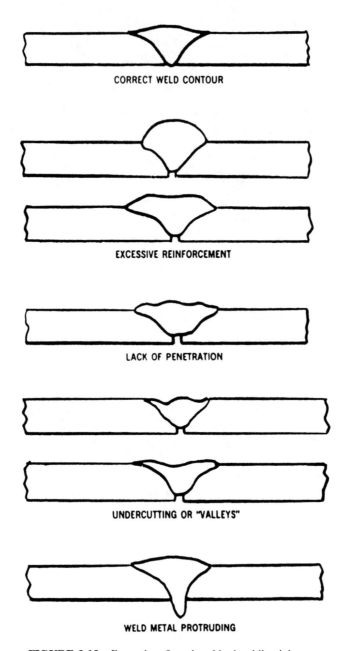

FIGURE 6.10   Examples of good and bad welding joints.

TROUBLESHOOTING

Some of the most common oxyacetylene-welding problems are summarized in Table 6.7. **Note:** Do not attempt to reweld a defective weld. Remove the defective weld metal from the joint and lay down a new weld bead. Rewelding commonly produces weak welds.

TABLE 6.7   Troubleshooting oxyacetylene welding.

Problem	Possible cause	Suggested remedy
Welding flame splits.	Dirty or clogged torch tip.	Stop welding, allow torch tip time to cool, thoroughly clean the tip, and then resume welding.
Sharp inner cone of flame disappears.	Dirty torch tip.	Stop welding, allow torch tip time to cool, thoroughly clean the tip, and then resume welding.
Premature ignition of gas mixture.	Sparks from weld puddle produce carbon deposits inside nozzle and on torch tip face.	Stop welding, allow torch tip time to cool, thoroughly clean the tip, and then resume welding.
Popping sound, or backfire (A single explosion or pop, or a series of small explosions occurring shortly after the torch is lit. Welding flame may disappear and then reappear, or remain extinguished).	a. Clogged torch tip. b. Pre-ignition of gas mixture in the torch tip, torch mixing chamber, or both. c. Overheated torch tip caused by holding tip to close to the work.	a. Stop welding, allow torch tip time to cool, clean the tip, and then resume welding. b. Increase oxygen and acetylene pressures slowly until popping sound is eliminated. c. Stop welding, allow torch tip time to cool, and then resume welding with tip held farther away from the work.
Squealing sound (Resulting from a very rapid series of small explosions after the torch is lit).	Same as "Popping sound."	Same as "Popping sound."
Flashback (Recession of flame into or in back of the torch mixing chamber).	Equipment problems such as loose or damaged torch tip, clogged torch tip orifices, kinked or damaged hoses; or incorrect welding procedures such as improper gas pressures, overheated torch tip, or failure to purge hose lines before lighting the torch.	First close the torch oxygen valve and then the acetylene valve. Wait to make sure the fire in the torch or hose has burned out. Check for damage to tip, regulators and hoses, and replace as necessary. Purge the hoses, check oxygen/acetylene pressures, and then relight torch using the standard lighting procedure.
Poor penetration.	Insufficient heat to penetrate to proper depth because torch tip to small.	Replace tip with one of appropriate size.
Holes burned in metal.	Excessive heat caused by tip too large for job.	Replace tip with one of appropriate size.
Uneven weld (Incomplete penetration. Good penetration in spots and partial penetration in between).	Torch is moved along the joint too quickly or slowly, zigzag movements not uniform nor in step with the puddle and zigzag movements overlapping.	Weld with a steady, uniform movement.
Fused portions at one side of the joint and not the other.	Caused by not moving the torch over the joint equally from side to side.	Move the torch so that it touches both sides of the joint uniformly.
Holes in the joint.	Caused by holding the flame too long in one place and overheating the metal.	Move the torch along the joint at a speed guaranteed not to overheat the base metal.
Holes in the joint at the end of the weld.	Failure to lift the torch and reduce the heat when reaching the end of the weld.	Use appropriate welding procedure.

(continued on next page)

TABLE 6.7   Troubleshooting oxyacetylene welding. *(continued)*

Problem	Possible cause	Suggested remedy
Oxide inclusions (Indicated by black specks on the broken surfaces of the weld).	Caused by not adequately cleaning the base metal surface prior to welding.	Make sure the base metal is as clean as possible before welding.
Adhesions	a. Insufficient heat was directed to one side of the joint. b. Welding speed too high causing the weld metal to break cleanly from the metal surface on one side.	a. Apply heat uniformly across joint. b. Decrease welding speed.
Brittle welds	Caused by using a carburizing flame. It is important for the welder to use the correct flame when welding.	Use a neutral flame on ferrous metals.
Overheating	Caused by moving the flame so slowly that too much heat is directed into the weld puddle. This results in the formation of excess metal "icicle" deposits on the bottom of the weld	Increase torch speed.
Welding (filler) rod sticks in weld puddle.	a. Failure to keep weld puddle molten. b. Failure to dip the welding rod into the weld puddle fast enough.	a. Maintain heat over weld puddle long enough to complete addition of welding rod. b. Dip welding rod into puddle more quickly.

MULTIPLE-CHOICE EXAM

1. Gas welding is a _____ welding process in which metal is heated to the liquid condition by the burning of gases with oxygen or other oxidizers.

   a. reactive   **b.** melting

   c. burning   **d.** fusion

2. Oxyacetylene welding is performed by the burning of two gases: oxygen and _____.

   a. acetylene   **b.** air

   c. hydrogen   **d.** nitrogen

3. When oxygen and acetylene are burned, the combination produces carbon dioxide and _____.

   a. water   **b.** sulfuric acid

   c. carbon monoxide   **d.** air

4. The most widely used gas in the welding industry is _____.

   a. carbon dioxide   **b.** hydrogen

   c. oxygen   **d.** acetylene

5. The pressure of stored acetylene is _____.

   a. woo psi   **b.** 225 psi

   c. 100 psi   **d.** 125 psi

6. Acetone is used in a tank of acetylene _____.

   a. to keep the acetylene from exploding

   b. to keep the acetylene gaseous

   c. to keep the acetylene in liquid form

   d. to make the tank of acetylene hard to transport

7. The shutoff valve is turned _____ times when on.

   a. one   **b.** one to one and one-half

   c. two   **d.** four

8.   Acetylene tanks are changed in an upright position _____.

   a. for safety purposes

   b. for stability of the gas

   c. to keep the gas from solidifying

   d. to make it easy to move around

9.   No torch or manifold should have an acetylene pressure above _____ psi for safety.

   a. 5   **b.** 10

   c. 15   **d.** 22

10. Oxygen is transported to the point of use in _____ form that is highly compressed in cylinders made for the purpose of containing the high pressure.

   a. free   **b.** solid

   c. liquid   **d.** gaseous

11. Other gases are supplied in the same type of cylinder as the oxygen cylinder. The gas inside is identified by the _____ of the tank.

   a. size   **b.** color

   c. weight   **d.** height

12. The pressure in the welding torch controlled for use _____.

   a. by a regulator   **b.** by a relief plug

   c. by a valve setup   **d.** by the atmosphere

13. There are _____ types of welding torches.

   a. five   **b.** two

   c. three   **d.** four

14. A _____ flame is used in welding.

   a. neutral   **b.** carburizing

   c. oxidizing   **d.** none of the above

15. In acetylene welding the tip is moved in a _____ motion, carrying the molten pool of metal across the work to be welded.

   a. circular   **b.** straight

   c. crescent   **d.** none of the above

16. To shut down after welding, the torch's _____ valve is closed first.

    a. oxygen

    b. acetylene

    c. air

    d. none of the above

17. When the torch is not being used for welding, it should be shut off and placed in its _____.

    a. clamp

    b. holder

    c. rack

    d. none of these

18. If you are unsure of whether or not you have a leak, it is best to:

    a. use the soap test

    b. use a match to see if it lights the gas

    c. use the striker to see if it ignites the gas

    d. go ahead and work as usual

19. If you are setting the pressure in the hoses, with the gas or oxygen flowing into the atmosphere, you must make certain:

    a. that there are no welders in the area

    b. that the tank valve is closed

    c. that there is no open flame or heat source in the area that could ignite the fuel

    d. that there is no leak

20. When the acetylene is first lit, the flame will be :

    a. yellow to orange in color

    b. a black, sooty color

    c. blue with a white tip

    d. white with orange

21. A(n) _____ flame should never be used for welding.

    a. oxidizing

    b. yellow

    c. blue

    d. sooty

22. When finished welding, you should turn off the _____ first.

    a. oxygen

    b. acetylene

    c. air

    d. tanks

23.   A pure acetylene flame contains no oxygen and is _____ in color.

    a. white   **b.** orange

    c. yellow   **d.** blue

24.   As oxygen is introduced to an acetylene flame, a cone begins to form, and the flame is called a _____ flame.

    a. carburizing   **b.** cutting

    c. welding   **d.** neutral

25.   When more oxygen is added to the acetylene flame, a _____ flame is the result.

    a. welding   **b.** cutting

    c. carburizing   **d.** neutral

26.   By adding more oxygen to a neutral flame, a _____ flame is produced.

    a. cutting   **b.** neutral

    c. carburizing   **d.** oxidizing

TRUE-FALSE EXAM

1.   Two-stage regulators have gauges that show working pressure in the lines and pressure in the tank.

    True   False

2.   Single-stage regulators have a tendency to freeze during cold weather.

    True   False

3.   There is no difference between the regulators used for oxygen and acetylene.

    True   False

4.   Probably the greatest disadvantage of using a single-stage regulator rather than a two-stage regulator is that a single-stage regulator will require adjustments to the torch as the pressure in the tank is lessened.

    True   False

5.   There is no major difference between the regulator used for oxygen and the one used for acetylene.

    True   False

6. A full tank of oxygen will contain approximately 2,200 psi of pressure, while a full acetylene tank will contain only 225 psi.

True   False

7. Regulators are the most important part of an oxyacetylene-welding unit.

True   False

8. Most regulators have signs not to oil them if they are used on the oxygen supply.

True   False

9. When regulators will not be used for a period of several weeks, you should relieve the pressure on the pressure-valve seat.

True   False

10. A well-maintained regulator will provide years of good service.

True   False

11. Frost on the regulator will affect its operation negatively.

True   False

12. Jump is caused by turning the acetylene gas higher at the torch.

True   False

13. Creep is probably the only true problem with regulators.

True   False

14. All hoses are colored red for acetylene and green for oxygen.

True   False

15. You should crack open the valve on the cylinder before installing a regulator.

True   False

16. Most welding hoses in use today are double, or Siamese, as they are called in the trade.

True   False

17. Older hoses can be checked for leaks by placing them in a pail of water and checking for bubbles when the hoses are under pressure.

True   False

18. Hose should not be stepped on while you are welding or at any time.

True   False

19. There are two types of torches available today: injector and high-pressure.

True   False

20. No welding blowpipe is complete without a welding tip.

True   False

21. Medium-pressure blowpipes are probably the most common type used for general- purpose welding.

True   False

22. Just as there are different sizes of welding tips, there are different sizes of cutting tips for the torch.

True   False

23. You can clean the tip of a welding torch by rubbing it across emery paper.

True   False

24. Flashback arrestors are attached to the hose end of the torch handle to eliminate the possibility of a flashback.

True   False

25. Newer types of torch lighters are made with a magneto, as was used on older-model automobiles to power spark plugs. Just squeeze, and a number of sparks are produced at the end of the lighter to ignite the gas.

True   False

MULTIPLE-CHOICE ANSWERS

1.	D	7.	B	12.	A	17.	C	22.	B
2.	A	8.	A	13.	B	18.	A	23.	C
3.	A	9.	C	14.	A	19.	C	24.	A
4.	D	10.	C	15.	C	20.	A	25.	D
5.	B	11.	B	16.	B	21.	A	26.	D
6.	A								

TRUE-FALSE ANSWERS

1.	T	8.	T	14.	T	20.	T
2.	T	9.	T	15.	T	21.	T
3.	F	10.	T	16.	T	22.	T
4.	T	11.	F	17.	T	23.	T
5.	F	12.	T	18.	T	24.	T
6.	T	13.	T	19.	F	25.	T
7.	T						

—NOTES—

Chapter 7

FLUX CORED ARC WELDING

The self-shielded flux-cored arc-welding process came about in the quest for automated welding. The process of putting a 9- or 18-inch electrode into a holder every few minutes takes time and effort on the part of the welder. If the electrode-replacement process could be eliminated, the time lost in changing electrodes would be better utilized in improving productivity.

One answer to the problem was to make the electrode of wire so it could be coiled or rolled up and fed out automatically or semiautomatically as needed. It would also be desirable inasmuch as it would allow more metal to be placed in the weld area and thus improve the time used to make a number of passes to accomplish the same result. By putting the flux inside the wire, it was available when needed and did not flake off or crack before it could be used. The outside cover of flux made it easy to handle, and the metal used for the electrode would be smooth on the outside and allow better guidance via a current-carrying guide tube. (See Figure 7.1.)

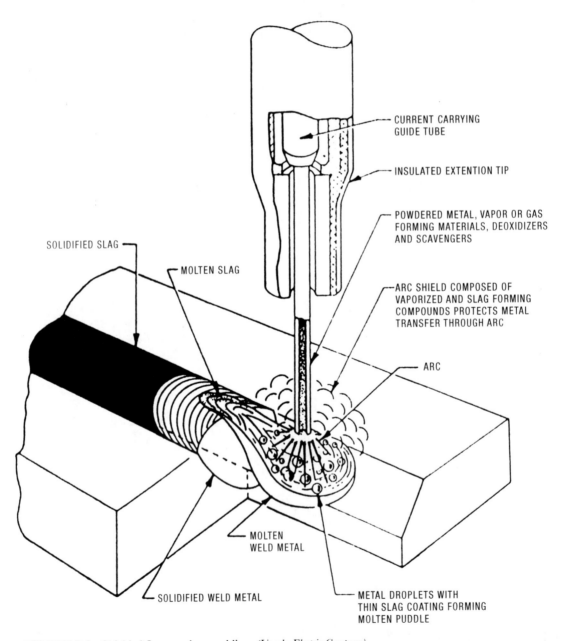

FIGURE 7.1   Shielded flux-cored arc welding. *(Lincoln Electric Company)*

The welder can activate the system by pressing a trigger to complete the welding circuit. With the semi-automatic gun he or she can reach into areas that are inaccessible to the semiautomatic equipment of other processes.

Advantages of the Flux Cored Arc Welding Process:

- When compared with stick-electrode welding, it gives deposition rates up to four times as great, often decreasing welding costs by as much as 50 to 75 percent.

- It eliminates the need for flux-handling and recovery equipment, as in submerged-arc welding, or for gas, gas-storage, piping, and metering equipment, as in gas-shielded mechanized welding. The semiautomatic process is applicable where other mechanized processes would be too unwieldy.

- It tolerates elements in steel that normally cause weld cracking when stick-electrode or one of the other mechanized welding processes are used.

- It produces crack-free welds in medium-carbon steel using normal welding procedures.

- Under normal conditions, it eliminates the problems of moisture pickup and storage that occur with low-hydrogen electrodes.

- It eliminates stub losses and the time that would be required to change electrodes with the stick-electrode process.

- It eliminates the need for wind shelters, required with gas-shielded welding in field erection, and permits fans and fast air-flow ventilation systems to be used for worker comfort in the shop.

- It permits more seams to be welded in one pass, saving welding time and the time that otherwise would be consumed in between-pass cleaning.

- It is adaptable to a variety of products and permits continuous operation at one welding station, even though a variety of assemblies with widely different joint requirements are run through it.

- It provides for fast filling of gouged-out voids, often required when making repairs to welds or steel castings.

- It gives the speed of mechanized welding to close quarters and reaches into spots inaccessible by other semiautomatic processes.

- It enables the bridging of gaps in fitup by allowing operator control of the penetration without reducing the quality of the weld, minimizing repair, rework, and rejects.

MULTIPLE-CHOICE EXAM

1. The self-shielded flux-cored welding process is an outgrowth of the _____ -metal-arc welding process.

    a. coded
    b. shielded
    c. fillet
    d. stressed

2. Where is the flux located in the flux-cored welding process?

    a. on the outside of the electrode
    b. in the middle of the electrode
    c. in the weld puddle
    d. in the arc it produces

3. How long is the electrode in flux-cored welding?

    a. 9 inches
    b. 18 inches
    c. an almost unlimited coil
    d. 30 inches

4. Why can flux-cored welding be used in industry so economically?

    a. It is easily adapted to automatic operation.
    b. It is semiautomatic in operation
    c. It is manually operated.
    d. It is electronically controlled.

5. The advantage of semiautomatic operation of the flux-cored process is:

    a. It can reach into areas that are inaccessible to other processes.
    b. It is simple to learn to operate.
    c. It is lightweight.
    d. It is less expensive than other processes.

6. One reason for incorporating the flux inside a tubular wire is to make feasible the coiling of the _____.

    a. flux
    b. electrode holder
    c. electrode
    d. cables

7.   What are the advantages of the flux-cored semiautomatic process over the stick-electrode?

    a. It has a higher deposition rate.

    b. It can use higher currents.

    c. It has automatic electrode feed.

    d. all of the above

    e. none of the above

8.   Although the American Welding Society calls it self-shielded flux-cored arc welding, it is often referred to as open-arc _____ welding.

    a. vapor   **b.** semi

    c. pulse   **d.** squirt

9.   In many shops the trade name for semiautomatic flux-cored arc welding is _____ welding.

    a. outer-shield   **b.** inner-shield

    c. shielded   **d.** arcless

10.   One of the advantages of flux-cored arc welding is its elimination of the need for handling _____.

    a. electrodes   **b.** slag

    c. flux   **d.** none of the above

11.   The flux-cored welding process eliminates the need for _____ shelters, required with gas-shielded welding in field erections, and permits fans and fast air-flow ven-tilation systems to be used for worker comfort in the shop.

    a. air   **b.** wind

    c. gas   **d.** arc

12.   The flux-cored welding process is adaptable to a variety of products, and it permits _____ operation at one welding station even though a variety of assemblies with widely different joint requirements are run through it.

    a. continuous   **b.** uneven

    c. even   **d.** none of the above

TRUE-FALSE EXAM

1. The self-shielded flux-cored arc-welding process is a progression of the shielded-metal arc-welding process.

    True   False

2. In the flux-cored welding process the flux is included in the electrode core.

    True   False

3. In flux-core welding the electrode is only 9 to 18 inches long.

    True   False

4. Flux-cored welding lends itself to semiautomatic operation very easily.

    True   False

5. The flux-cored welding process makes use of its ability to handle high currents.

    True   False

6. Higher deposition rates are an advantage of flux-cored welding.

    True   False

7. The accepted term for flux-cored welding is squirt in the shop.

    True   False

8. The trade name Outer-Shield is used in many shops to identify the flux-core process.

    True   False

9. Before all-position electrodes for self-shielded flux-cored arc welding were developed, beam web-to-column connections in building erection were made with stick electrodes.

    True   False

10. When compared with stick-electrode welding, flux-cored welding gives deposition rates up to 10 times as great.

    True   False

11. The flux-cored process eliminates the need for flux handling and recovery equipment.

    True   False

12. The flux-cored process has tolerance for elements in steel that normally cause weld cracking when stick-electrode or one of the other mechanized welding processes is used.

    True   False

MULTIPLE-CHOICE ANSWERS

1.	B	4.	B	7.	D	9.	B	11.	B
2.	B	5.	A	8.	D	10.	C	12.	A
3.	C	6.	C						

TRUE-FALSE ANSWERS

1.	T	4.	T	7.	T	10.	F
2.	T	5.	T	8.	F	11.	T
3.	F	6.	T	9.	T	12.	T

—NOTES—

Chapter 8

GAS METAL
ARC WELDING

Gas-metal-arc welding (GMAW) is fast and economical, because there is no frequent changing of electrodes as with stick-type electrode welding. And there is no slag formed over the weld. The process often can be automated and, if done manually, the welding head is light and compact. GMAW is a logical outgrowth of gas-tungsten-arc welding. It differs in that the arc is maintained between an automatically fed, consumable wire electrode and the work-piece. It automatically provides the additional filler. It was formerly referred to as MIG (metal-inert-gas) welding. Argon, helium, or mixtures of the two can be used for welding virtually any metal. They are used primarily for welding nonferrous metals. In welding steel, some O_2 or CO_2 is added to improve the arc stability. It also reduces weld spatter. The cheaper CO_2 alone can be used for welding steel, provided that a deoxidizing electrode wire is used.

Shielding gases have a considerable effect on the nature of the metal (drop) transfer from the electrode to the work. They also affect the tendency for undercutting. Electronic controls alter the waveform of the current to make it possible to vary the mechanism of metal transfer-by drop spray or shot-circuiting drops.

Some of these variations in the basic process are:

- Pulsed-arc welding

- Short-circuiting arc welding

- Spray-transfer welding

- Buried arc welding (carbon-dioxide gas is used and the arc is buried in its own crater)

Gas-metal-arc welding was originally developed to solve the problem with TIG welding, which could only weld metals up to 1/4 inch thick. GMAW or MIG welding can weld metal more than 1/4 inch thick. One way in which MIG varies from TIG is that the MIG has an electrode that disappears into the weld puddle, thus producing a strong, clean weld.

Shielding gas for the MIG process is typically:

- Carbon dioxide

- Argon/oxygen

- Argon/carbon dioxide

The finished weld has no slag and virtually no spatter. This process can be used on a wide variety of material types including:

- Mild steel

- Low-alloy steel

- Aluminum

- Stainless steel

This type of welding is semiautomatic. It uses a handheld gun to which the electrode is fed automatically, or it is used as a full automatic process. The welding guns or heads are similar to those used with gas-shielded flux-cored welding. When the term "manual" is used, it usually refers to the semiautomatic process that utilizes the hand-held gun.

MULTIPLE-CHOICE EXAM

1. Another name for gas-metal-arc welding is:

   a. tungsten   **b.** submerged

   c. MIG   **d.** TIG

2. The following gases may be used for proper gas-metal-arc welding operation.

   a. helium, carbon dioxide

   b. helium, argon, carbon dioxide

   c. air, oxygen, carbon dioxide

   d. oxygen, argon

3. Although the gas shield is effective in shielding the molten metal from the air, _____ are usually added as alloys in the electrode.

   a. deoxidizers   **b.** oxidizers

   c. chemicals   **d.** none of these

4. When the term manual gas-metal-arc welding is used, the _____ process with its hand-held gun is implied.

   a. automatic   **b.** semiautomatic

   c. both of these   **d.** neither of these

5. Which technique of welding is meant by the term short welding?

   a. shielded welding   **b.** gas welding

   c. spray-arc welding   **d.** short-circuiting welding

6. To use short-arc welding efficiently, special power sources with adjustable slope, voltage, and _____ characteristics are required.

   a. resistance   **b.** inductance

   c. capacitance   **d.** current

7. Which type of welding is a logical outgrowth of gas-tungsten-arc welding?

   a. pressure-gas welding   **b.** arc welding

   c. gas-tungsten-arc welding   **d.** flux-cored-arc welding

8. Gas-metal-arc welding was originally developed to solve the problem of welding metals thicker than
_____ inch using the TIG method.

    a. 1/2 inch   **b.** 1/4 inch

    c. 1 inch   **d.** 1 1/2 inches

9. In the gas-metal-arc-welding process the electrode is consumed; however, in the _____ welding process the electrode is not consumed.

    a. DC arc-   **b.** tungsten

    c. TIG   **d.** MIG

10. MIG welding produces a:

    a. glob of weld fill with lots of slag

    b. spotty, slag-covered weld

    c. poor weld but clean

    d. strong, clean weld

TRUE-FALSE EXAM

1. Gas-metal-arc welding was originally referred to as MIG welding.

    True   False

2. The shielding gases can be oxygen and carbon dioxide.

    True   False

3. CO_2 alone can be used for welding steel with MIG.

    True   False

4. Buried-arc welding is another variation in which carbon-dioxide-rich gas is used and the arc is buried in its own crater.

    True   False

5. The shielding gases have no effect on the nature of the metal transfer from the electrode to the work.
   True   False

6. Gas-metal-arc welding is fast and economical.
   True   False

7. MIG welding may be used with all the major commercial metals, including carbon, alloy, and stainless steels as well as aluminum, magnesium, copper, iron, titanium, and zirconium.
   True   False

8. MIG welding may be completely or fully automatic.
   True   False

9. Short-arc welding uses low currents, low voltages, and small-diameter wires.
   True   False

10. The power sources for short-arc welding do not produce reliable voltages and currents for good welds.
    True   False

11. The short-arc welding technique produces a high amount of heat.
    True   False

12. Short-arc welding tolerates poor fit-up and permits the bridging of wide gaps.
    True   False

13. There is a process of welding that is referred to as globular.
    True   False

14. Spray-arc MIG welding produces an intensely hot, higher-voltage arc and gives a higher disposition rate than short-arc welding.
    True   False

15. The spray arc technique is not recommended for 1/8 inch and thicker metals.
    True   False

MULTIPLE-CHOICE ANSWERS

1. C	3. A	5. D	7. C	9. C
2. B	4. B	6. B	8. B	10. D

TRUE-FALSE ANSWERS

1. T	5. F	9. T	13. T
2. T	6. T	10. F	14. T
3. T	7. T	11. F	15. F
4. T	8. T	12. T	

Chapter 9
GAS TUNGSTEN ARC WELDING

The American Welding Society's definition of gas tungsten arc (TIG) welding is "an arc-welding process wherein coalescence is produced by heating with an arc between a tungsten electrode and the work." A filler metal may or may not be used. Shielding is obtained with a gas or a gun mixture. TIG welding was developed in the 1940s for welding magnesium, aluminum, and stainless steel. It is widely used in the aircraft industry. TIG uses a tungsten electrode held in a special holder through which an inert gas is supplied with a sufficient flow to form an inert shield around the arc and the molten pool of metal, thus shielding them from the atmosphere. This process uses argon, helium, or an argon/helium mixture for the shielding. Since the tungsten electrodes often are treated with thorium or zirconium, they provide better current-carrying and electron emission. A high-frequency, high-voltage current usually is superimposed on the regular AC or DC welding current to make it easier to start and maintain the arc. The tungsten electrode is virtually not consumed at arc temperatures in the inert gases, and the arc length remains constant so that the arc is stable and easy to maintain. A clean, strong weld is thus produced without slag.

If filler is needed, it must be supplied by a separate wire, as in gas-flame welding. However in many applications where a close fit exists between the parts being welded, no filler metal may be needed. Continuous fine filler is heated by the passage of an ac current so that it melts as it feeds into the weld puddle just behind the arc. This is as a result of the I^2R effect. The deposition rate is several times what can be achieved with a cold wire, and it can be increased further by oscillating the filler wire from side to side when making a wide weld. The hot-wire process cannot be used in the welding of copper or aluminum. This is due to their low resistivities.

Gas-tungsten-arc welding produces very clean welds, and no special cleaning or slag removal is required because no flux is used. With skillful operators, welds that are scarcely visible often can be made. However, the surfaces to be welded must be clean and free of oil, grease, paint, or rust, because the inert gas does not provide any cleaning or fluxing action.

SPOT WELDING

A variation of the gas-tungsten-arc-welding process is used for making spot welds. The two pieces of metal do not necessarily need to be accessible on both sides of the joint. The basic procedure is shown in Figure 9.1. If this process a modified, vented inert-gas, a tungsten-arc gun, and a nozzle are used. The nozzle is pressed against one of the two pieces of the joint. The work pieces must be sufficiently rigid to sustain the pressure

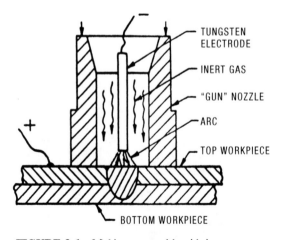

FIGURE 9.1 Making spot welds with inert-gas shielded tungsten arc process.

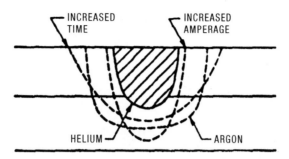

FIGURE 9.2 How changes in time, current and shielding gas affect the shape of the weld nugget.

that is applied to one side to hold them in reasonably good contact. The arc between the tungsten electrode and the upper work piece provides the necessary heat. The inert gas, usually argon or helium, flows through the nozzle and provides a shielding atmosphere. An automatic control moves the electrode to make momentary contact with the work piece to start the arc. It then withdraws and holds it at a correct distance to maintain the arc. The duration of the arc is timed automatically so that the two work pieces are heated sufficiently to form a spot weld under the pressure of the gun nozzle. The depth and size of the weld nugget are controlled by the amperage, the time, and the type of shielding gas. (See Figure 9.2.)

Because access to only one side of the work is required, this type of spot welding has an advantage over resistance spot welding especially in certain applications, as in fastening relatively thin sheet metal to a heavier framework.

MULTIPLE-CHOICE EXAM

1.   Tungsten-inert-gas welding is called _____ in the trade.
   a. MIGW   **b.** TIGW
   c. MIG   **d.** TIG

2.   What is the result of the atmospheric contaminants becoming part of the welding process?
   a. clean joints   **b.** strong joints
   c. joints with nice lines   **d.** weak spots in the joint

3.   An inert gas is:
   a. active   **b.** inactive
   c. lazy   **d.** not reactive with other gases and metals

4. Tungsten-inert-gas welding was originally developed specifically for welding manganese, aluminum, and _____.

   a. stainless steel   **b.** ferrous metals

   c. heavy metals   **d.** copper

5. Before TIG welding was introduced, welders had to resort to using _____ to remove contaminants from the weld.

   a. electrodes   **b.** heat

   c. skill   **d.** flux

6. Tungsten is used for the electrode in TIG welding because:

   a. it is plentiful   **b.** it is inexpensive

   c. it does not melt easily   **d.** it melts easily

7. The TIG process is particularly suitable for welding _____ materials where the requirements for quality and finish are exacting.

   a. heavy   **b.** thick

   c. thin   **d.** rough

8. The TIG electrode that does not melt in the arc is made of:

   a. aluminum   **b.** copper

   c. tungsten   **d.** magnesium

9. The advantage of using a hot wire with oscillation for filler in TIG welding is that:

   a. it has a less rapid deposition rate

   b. it has a more rapid deposition rate

   c. it has a very easy-to-use coating

   d. it can be easily coiled

10. Gas-tungsten-arc spot welding can be used where only one side of the work piece is available for applying _____.

   a. pressure   **b.** heat

   c. cooling water   **d.** the arc

TRUE-FALSE EXAM

1. The advantage of a spot-welding process such as gas tungsten is that it can be used when access to only one side of the joint is available.

    True   False

2. Argon is used in the TIG-welding process.

    True   False

3. Helium is used tn the TIG-welding process.

    True   False

4. Carbon dioxide is used in TIG welding.

    True   False

5. The electrode in the TIG process is melted in the arc.

    True   False

6. Gas-tungsten-arc welding produces very clean welds, and no special cleaning or slag removal is required.

    True   False

7. Tungsten electrodes often are treated with thorium or zirconium to provide better current-carrying and electron-emission characteristics.

    True   False

8. The abbreviation often used to indicate gas-tungsten-arc welding is GTAW.

    True   False

9. One of the main advantages of TIG welding is that it produces a clean, strong weld.

    True   False

10. TIG welding is not easily adapted to automatic welding.

    True   False

MULTIPLE-CHOICE ANSWERS

1. D	3. D	5. D	7. C	9. B
2. D	4. A	6. C	8. C	10. A

TRUE-FALSE ANSWERS

1. T	4. F	7. T	9. T
2. T	5. F	8. T	10. F
3. T	6. T		

Chapter 10

BRAZE WELDING

Braze welding and brazing are two different processes. They differ in two ways. Braze welding does not use capillary attraction to distribute the filler material. Molten filler metal is deposited by gravity. Because relatively low temperatures are required, there is less danger of causing warping than might result if arc welding or oxyacetylene welding were used. It is most effective for the repair of steel parts and ferrous castings. The strength obtained is determined by the braze metal used. Considerable buildup of the braze material is required if full strength of the repaired part is required.

Nearly all braze welding is done with an oxyacetylene torch. The surfaces should be tinned with a thin coating of the brazing metal before the remainder of the filler metal is added.

The design of the joints being secured makes braze welding different from brazing. Braze welding is used for groove, fillet, and other weld joints where the filler material does not take advantage of capillary action. Braze welding uses a nonferrous filler metal that has a melting point below that of the parent material but above 800 degrees F.

The technique for braze welding is similar to that used for fusion welding. The important exception is that the parent metal is not melted but only raised to the tinning temperature.

FILLER MATERIAL

Both brazing and braze welding use a nonferrous filler material. In most cases this will be an alloy of copper, zinc, and tin. Two commonly used copper alloys for brazing and braze welding are brass and bronze. Brass is an alloy consisting of copper and zinc. Bronze is composed mainly of copper and tin. More popular for both joining techniques is brass. It is interesting to note that even experienced welders call brazing and braze-welding rods bronze rods when actually they are using brass rods consisting of copper, zinc, and about 1 percent tin.

There are other brazing and braze-welding rods in addition to the copper alloys that are used for joining special metals. These include:

- nickel and chromium alloys

- silver alloys

- copper and gold alloys

- aluminum and silicon alloys

- magnesium alloys

The do-it-yourself welder will probably use a bronze rod for most types of brazing and braze welding. Bronze is a good nonferrous filler material for joining cast iron and steel.

FLUX

Flux plays a very important role in braze welding and brazing. Most brazing processes require flux to prevent oxidation during the application of heat. A flux is used in soldering to protect the metal surfaces being joined from atmospheric contaminants. In brazing and braze welding, flux is used for the same reason. Flux aids in capillary attraction of nonferrous metals and in dissolving oxides that form during either brazing or braze welding.

Flux is available in powder, paste, or liquid form. Flux-coated bronze rods are also available. There is no such thing as all-purpose flux. Many fluxes are designed for specific applications. Most fluxes are sold chemically pure. Note that there will be some mention of this fact on the label.

Borax is an old standby flux that has been around for many years. It can be successfully used for brazing brass, bronze, steel, and malleable iron. One of the most popular mixtures consists of 75 percent borax and 25 percent boric acid in paste form. This anti-borax flux, as it is sometimes called, is suitable for brazing those metals mentioned previously.

Most fluxes are sold in powder form. This is quite convenient, as the flux can then be applied to the metal in two different ways. The first way is to heat up the end of a bronze rod and dip it into the powdered flux. The hot rod will cause the flux to adhere to the rod, and the rod can then be used as is, with a coating of flux.

Powdered fluxes can also be made into paste form. This type of flux is usually made by adding water until the right consistency is achieved. Paste fluxes are commonly applied to the metal with a small brush. It is important to read the container label. Some fluxes require mixing with alcohol instead of water to form a paste.

MULTIPLE-CHOICE EXAM

1. What are the two major differences between brazing and braze welding?

    a. the joints between the pieces and the use or not of capillary action

    b. the joints are basically the same

    c. the metals used for the rod and the fluxes used

    d. none of the above

2. Braze welding gives better joints without having to have close _____.

    a. tolerances   **b.** joints

    c. fluxing   **d.** none of the above

3. When the filler metal comes in contact with the hot parent metal, it tries to open up the _____.

    a. distances between them   **b.** grain

    c. flux   **d.** none of the above

4. One of the most popular joints used in braze welding is the _____ joint.

    a. illet   **b.** butt

    c. scarf   **d.** ap

5. When joining metal that is thicker than ____ inch, it is necessary to bevel the edges of the joint.

    a. 3/4   **b.** 1

    c. 1/4   **d.** 1/2

6.   Besides solvent cleaning the metal to be joined can also be _____.

     a. scalded   **b.** scraped

     c. torched   **d.** flame-cleaned

7.   After the pieces to be joined have been cleaned, they must be _____ and held in place during the braze-welding procedure.

     a. aligned   **b.** ground smooth

     c. sanded   **d.** none of the above

     e. all of the above

8.   What is often needed in braze welding to reduce the possibility of expansion and contraction stresses?

     a. flux   **b.** a very hot flame

     c. preheating   **d.** none of the above

     e. all of the above

9.   Preheating is usually only necessary with the thicker pieces of metal over:

     a. 1 inch thick   **b.** 3/4 inch thick

     c. 1/2 inch thick   **d.** 1/4 inch thick

10.   As the parent metal becomes hot enough to melt the bronze rod, it will form a thin coat on the metal surfaces. This is known as:

     a. tinning   **b.** fluxing

     c. rose glow   **d.** soldering

     e. none of the above

TRUE-FALSE EXAM

1. Joint design for braze welding is similar to those joints used in fusion welding.
    True   False

2. Braze welding is a popular means of joining many types of metal using the butt joint.
    True   False

3. When joining metal over 1/4" thick it should have beveled edges for the joint.
    True   False

4. All parts of the metal to be joined must be clean of paint, grease, oil or any other foreign material.
    True   False

5. Another method used to clean metal of surface dirt is called flame cleaning.
    True   False

6. Preheating is not often required in braze welding.
    True   False

7. It is not important when braze welding to keep the torch flame moving over the surface of the joints.
    True   False

8. While heating the joint surface, the end of the bronze welding rod should also be heated and then dipped into the powdered flux.
    True   False

9. A thin coating on the metal surfaces put there by heating the welding rod is called tinning.
    True   False

10. Even cooling is not important when braze welding.
    True   False

MULTIPLE-CHOICE ANSWERS

1.	A	3.	B	5.	C	7.	A	9.	D
2.	A	4.	B	6.	D	8.	C	10.	A

TRUE-FALSE ANSWERS

1.	T	4.	T	7.	F	9.	T
2.	T	5.	T	8.	T	10.	F
3.	T	6.	F				

METALLURGY, BRAZING, AND SOLDERING

—NOTES—

Chapter 11
BRAZING

Brazing is also known as hard soldering. Metal clearances and the bead are similar to oxyacetylene welding of steel. However, a large amount of filler metal is used.

Brazing differs from other welding processes in that a nonferrous filler metal is used. The filler melts at a temperature above 450 degrees C (840 degrees F). This is below the temperature of the base metal. Coalescence is between the filler and the base metals or between the base metals and alloys that may be formed. The result is that the base metals are not melted. Capillary attraction plays an important role in distributing the filler metal in the joint.

Some important characteristics and requirements of brazing are:

- The 450-degree C (840-degree F) temperature is an arbitrary one, used to distinguish brazing from soldering. Because the base metals are not melted, the brazing metal always is different in composition from the base metals.

- For capillary attraction to exist, the clearance between the parts being joined must be quite small.

- In order for capillary attraction to be effective, the base-metal surfaces must be clean so that the brazing metal can easily wet them. This usually requires the use of a flux or inert atmospheres in a vacuum furnace.

- Heating a brazed joint above the melting point of the brazing metal, even if followed by subsequent cooling, may destroy the integrity of the joint.

Advantages of brazing include:

- Virtually all metals can be joined by some type of brazing metals.

- Less heating is required than for welding

- The process can be done more quickly and more economically than most others.

- Lower temperatures are used.

- There are fewer difficulties due to distortion.

- Thinner and more complex assemblies can be successfully joined.

- Many brazing operations can be mechanized.

Coalescence is the growing together or growth into one body of the base metal parts.
These advantages makes it apparent that brazing is well suited for use in mass production and for delicate assemblies.

Disadvantages of brazing include:

- Reheating can cause inadvertent melting of the braze metal.

- Bronze metal running out of the joint weakens or destroys the joint. Too often this occurs when people apply heat to brazed parts. (Not a defect of brazing, but it can lead to most unfortunate results).

- Brazing specified for use in products that later might be subjected to abuse, adequate warning should be given to those who use the brazed device

STRENGTH

A brazed joint obtains its strength from a combination of the braze metal and the base-metal alloy that is formed. Penetration of the low-viscosity brazing metal into the grain boundaries of the base metal also adds strength. The strength of a properly brazed joint is between that of the base metal and the braze metal. However, the strength also is a function of the clearance between the parts being brazed. There must be sufficient clearance so that the braze metal will wet the joint and flow into it. Beyond this amount of clearance, the strength decreases rapidly down to that of the braze metal. The proper amount of clearance varies considerably, depending primarily on the type of braze metal. Copper requires virtually no clearance when heated in a hydrogen atmosphere.

A slight press fit of about 0.1 percent per unit of diameter (0.001 inch per inch) is recommended. For silver-alloy brazing metals the absolute clearance is about 0.04 to 0.05 mm (0.0015 to 0.002 inch).

When 60-40 brass is used for brazing iron and copper, a clearance of about 0.52 to 0.76 mm (0.02 to 0.03 inch) is desirable.

METALS

The most commonly used brazing metals are:

- Brazing brass

- Copper silicon

- Copper phosphorus

- Nickel silver

- Aluminum alloys

- Silver alloys

- Manganese bronze

Table 11.1 lists some of the more frequently used brazing metals and their usages. Table 11.2 shows welding-tip sizes and gas pressures most often used.

Copper is used only for brazing steel and other high-melting-point alloys, such as high-speed steel and tungsten carbide. Its use is confined almost exclusively to furnace heating in a protective hydrogen atmosphere in which the copper is extremely fluid and requires no flux.

Copper brazing is used extensively for assemblies composed of low-carbon-steel stampings, screw-machine parts, and tubing, such as are common in mass-produced products. Copper alloys, in the form of "spelter" brass, were the earliest brazing materials. The most common uses today for copper alloys in brazing are in the repair of steel and iron castings. Tobin and manganese bronzes are frequently used for this purpose. Phos-copper is used extensively in brazing copper, but should not be used on ferrous alloys. Silver alloys are widely used in fabricating copper and nickel alloys. Although these brazing alloys are expensive, such a small amount is required that the cost per joint is low. One typical silver-phosphorus-copper alloy, known as Sil-Fos®, is self-fluxing on clean copper and is used extensively for brazing this material. Silver alloys also are used for brazing stainless steel. However, because the brazing temperatures are in the range of carbide precipitation, only stabilized stainless steels should be brazed with these alloys if continued corrosion resistance is desired.

TABLE 11.1 Brazing metals and their uses.

Braze Metal	Composition	Brazing Process	Base Metals
Brazing brass	60% Cu, 40% Zn	Torch Furnace Dip Flow	Steel, copper, high copper alloys, nickel, nickel alloys, stainless steel
Manganese bronze	58.5% Cu, 1% Sn, 1% Fe, 0.25% Mn, 39.5% Zn	Torch	Steel, copper, high copper alloys, nickel, nickel alloys, stainless steel
Nickel silver	18% Ni, 55–65% Cu, 27–17% Zn	Torch Induction	Steel, nickel, nickel alloys
Copper silicon	1.5% Si, 0.25% Mn, 98.25% Cu, 1.5% Si, 1.00% Zn, 97.5% Cu	Torch	Steel
Silver alloys (no phosphorus)	5–80% Ag, 15–52% Cu, balance Zn + Sn + Cd	Torch Furnace Induction Resistance Dip	Steel, copper, copper alloys, nickel, nickel alloys, stainless steel
Silver alloys (with phosphorus)	15% Ag, 5% P, 80% Cu	Torch Furnace Induction Resistance Dip	Copper, copper alloys
Copper phosphorus	93% Cu, 7% P	Torch Furnace Induction Resistance	Copper, copper alloys

Aluminum-silicon alloys containing about 6 to 12 percent silicon are used for brazing aluminum and aluminum alloys. By using a braze metal that is not greatly unlike the base metal, the possibility of galvanic corrosion is reduced. However, because these brazing alloys have melting points of about 610 degrees C (1130 degrees F) and the melting temperature of commonly brazed aluminum alloys, such as 3003, is around 669 degrees C (1290 degrees F), control of the temperature used in brazing is quite critical. Thus, in brazing aluminum, good temperature control must be exercised, and proper fluxing action, surface cleaning, and/or

TABLE 11.2 Welding tip sizes and pressure settings for brazing.

Metal Thickness (inches)	Tip Size (inches)	Size of Brazing Rod (inches)	Oxygen Pressure (psi)	Acetylene Pressure (psi)
1/32	1	1/16	5	5
3/64	2	1/16	5	5
1/16	3	1/16	5	5
3/32	4	3/32	5	5
1/8	5	3/32	5	5
3/16	6	3/32	6	6
1/4	7	1/8	7	7
5/16	8	5/32	8	8

use of a controlled-atmosphere or vacuum environment must be utilized to assure adequate flow of the braze metal and avoid damage to the base metal.

A commonly used procedure in connection with brazing aluminum is to use sheets that have one or both surfaces coated with the brazing alloy. The thickness of the coat is about 10 percent of the total sheet thickness. These "brazing sheets" have sufficient coating to form adequate fillets. Joints are made merely by coating the joint area with suitable flux followed by heating.

FLUXES

Brazing fluxes play a very important part in brazing by:

- Dissolving oxides that may be on the surface prior to heating.

- Preventing the formation of oxides during heating.

- Lowering the surface tension of the molten brazing metal and thus promoting its flow into the joint.

Cleanliness is the primary factor affecting the quality and uniformity of brazed joints. Although fluxes will dissolve modest amounts of oxides, they are not cleaners. Before a flux is applied, dirt, particularly oil, should be removed from the surfaces that are to be brazed. The less the flux has to do prior to heating, the more effective it will be during heating.

Borax was a commonly used brazing flux. Fused borax should be used because the water in ordinary borax causes bubbling when heat is applied to the flux. Alcohol can be mixed with fused borax to form a paste. Many modern fluxes are available that have lower melting temperatures than borax and are somewhat more effective in removing oxidation.

Fluxes should be selected with reference to the base metal. Paste fluxes usually are used for furnace, induction, and dip brazing, and either paste or powdered fluxes are used for torch brazing. In furnace, induction, and dip brazing the flux ordinarily is brushed onto the surfaces. In torch brazing it often applied by dipping the heated end of the filler wire into the flux.

Fluxes for aluminum usually are mixtures of metallic hallide salts. The base typically is potassium chloride (from 15 to 85 percent of the flux). Activators, such as fluorides or lithium compounds, are added. These fluxes do not dissolve the surface oxide film on aluminum.

Most brazing fluxes are corrosive. The residue should be removed from the work after the brazing is completed. This is particularly important in the case of aluminum. Much effort has been devoted to developing fluxless procedures for brazing aluminum.

BRAZING TECHNIQUES

Application of Brazing Metal

Brazing metal is applied to joints in three ways. The oldest, and a common method used in torch brazing, uses the brazing metal in the form of a rod or wire. When the joint has been heated to a sufficient temperature so that the base metal will melt, the brazing wire or rod is then melted by the torch, and capillary attraction causes it to flow into the joint. Although the base metal should be hot enough to melt the braze metal and assure that it remains molten and flows into the joint, the actual melting should be done with the torch.

Heating Methods

A common source of heat for brazing is a gas-flame torch. In this torch-brazing procedure, oxyacetylene, oxyhydrogen, or other gas-flame sources can be used. Most repair brazing is done in this manner because of its flexibility and simplicity, but the process also is widely used in production brazing, Its major drawbacks are the difficulty in obtaining uniform heating and proper control of the temperature, and the requirement for costly skilled labor. In production-type torch brazing, specially shaped torches often are used to speed the heating and to aid in reducing the amount of skill required.

Large amounts of brazing are done in controlled-atmosphere furnaces. In such furnace brazing, the brazing metal must be preloaded into the work. If the work is not of such a nature that its preassembly will hold the parts in proper alignment and with adequate pressure, brazing jigs or fixtures must be used. Assemblies that are to be brazed usually can be designed so that such jigs or fixtures will not be needed if adequate consideration is given to this matter; often a light press fit will suffice.

Flux Removal

Although not all brazing fluxes are corrosive, most of them are. Consequently, flux residues usually must be completely removed. Most of the commonly used fluxes are soluble in hot water, so their removal is not difficult. In most cases immersion in a tank of hot water for a few minutes will give satisfactory results, provided that the water is kept really hot. Usually it is better to remove the flux residue while it is still hot. Blasting with sand or grit is also an effective method of flux removal, but this procedure cannot be utilized if the surface finish must be maintained. Such drastic treatment seldom is necessary.

Fluxless Brazing

Both the application of brazing flux and the removal of flux residues involve significant costs, particularly where complex joints and assemblies are involved. Consequently, a large amount of work has been devoted to the development of procedures whereby flux is not required, particularly for brazing aluminum. This work has been spurred by the obvious advantages of aluminum as a lightweight and excellent heat conductor for use in heat-transfer applications such as radiators in automobiles, where weight reduction is of increasing importance.

The brazing of aluminum is complicated by:

- The high refractory oxide surface film
- Its low melting point
- Its high galvanic potential

However, the successful fluxless brazing of aluminum has been achieved by employing rather complicated vacuum-furnace techniques, utilizing vacuums up to 0.0013 Pa (1×10^{-5} torr). Often a "getter" metal is employed to aid in absorbing the small amount of oxygen, nitrogen, and other occluded gases that remain in the "vacuum" or that may be evolved from the aluminum being brazed. The aluminum must be carefully degreased prior to brazing, and the design of the joint is quite critical. Sharp V-edge joints appear to give the best results.

Some success has been achieved in the fluxless induction brazing of aluminum in air. Induction-heating coils are designed to act as clamps to hold the pieces being brazed together under pressure, and an aluminum braze metal containing about 7 percent silicon and 2.5 percent magnesium is used. The resulting magnesium vapor apparently reduces some of the oxide on the surface of the aluminum and thus permits the braze metal to flow and cover the aluminum surface.

Brazed Joints

Three types of brazed joints are used:

- Butt

- Scarf

- Lap or shear

These, together with some examples of good and poor design details, are shown in Figure 11.1.

Because the basic strength of a brazed joint is somewhat less than that of the parent metals, the desired strength must be obtained by using sufficient joint area. This means using some type of lap joint where maximum strength is required. If joints are made very carefully, a lap of 1 or 1 1/4 times the thickness of the metal can develop strength as great as that of the parent metal. However, for joints that are to be made in routine production, it is best to use a lap equal to 3 times the material thickness. Full electrical conductivity usually can be obtained with a lap about 1 1/2 times the material thickness.

If maximum joint strength is desired, it is important to have some pressure applied to the parts during heating and until the braze metal has cooled sufficiently to attain most of its strength. In many cases, the needed pressure can be obtained automatically through proper joint selection and design.

In designing joints that are to be brazed, one must make sure that no gases can be trapped within the joint. Trapped gas may prevent the filler metal from flowing throughout the joint, owing to pressure developed during heating.

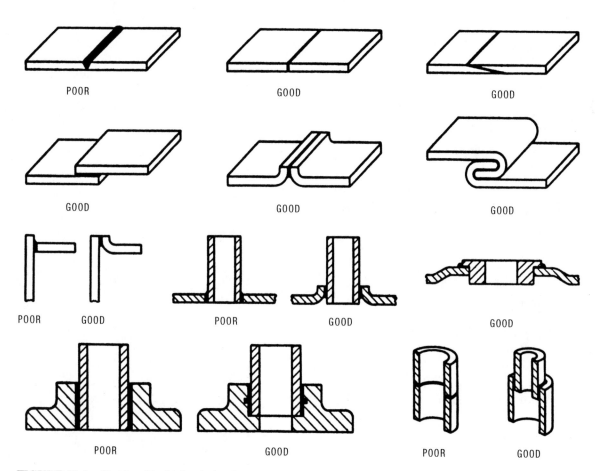

FIGURE 11.1 Good and bad joint design for brazing.

MULTIPLE-CHOICE EXAM

1. When brazing metals, the fit between the pieces is very important and must be _____.
    a. of the same material   **b.** clamped
    c. loose   **d.** tight

2. The flowing of filler material in brazing is known as:
    a. capillary attraction   **b.** capillary rejection
    c. cold flow   **d.** hot flow

3. A brazed joint is stronger than a _____ joint.
    a. soldered   **b.** welded
    c. bonded   **d.** glued

4. The two most commonly used copper alloys for brazing are brass and _____.
    a. silver   **b.** gold
    c. copper dioxide   **d.** bronze

5. What plays an important role in keeping a brazed joint free of contamination?
    a. hydrogen   **b.** nitrogen
    c. flux   **d.** heat

6. Most fluxes are sold in _____ form.
    a. paste   **b.** powder
    c. liquid   **d.** none of the above

7. A joint clearance of 0.002 to _____ inch is acceptable when brazing copper or silver alloys.
    a. 0.040   **b.** 0.006
    c. 0.009   **d.** 1.00

8. Metals to be brazed must be cleaned with a wire brush if there is _____ on the surface.
    a. rust   **b.** flux
    c. oil   **d.** wax

9. When brazing, it is important to choose the right _____ for the torch.

   a. gas

   b. tip

   c. hose

   d. rod

10. Which type of flame should be used for brazing?

    a. orange-blue

    b. neutral

    c. blue

    d. full

11. The melting point of all copper-alloy brazing rods, both brass and bronze, is between 1,000 and _____ degrees F.

    a. 1200

    b. 1400

    c. 1600

    d. 1800

12. When heated properly, the flux will boil at about _____ degrees F.

    a. 100

    b. 200

    c. 212

    d. 300

13. Once the filler material flows over the heated area of the joint in brazing, _____ will take place.

    a. bonding

    b. boiling

    c. fluxing

    d. coalescence

14. It is important to keep in mind that both pieces of metal must be of _____ temperature, because the filler rod will flow only to the hotter piece.

    a. a lower

    b. a higher

    c. unequal

    d. equal

15. The obvious main advantage of brazing over fusion welding is that it is the best way to join _____ metals.

    a. like

    b. similar

    c. plasticized

    d. dissimilar

16. Just about any two metals can be joined by brazing, with the exception of aluminum and _____.

    a. copper

    b. zinc

    c. babbit

    d. magnesium

17.   Aluminum can be brazed in the same manner as silver and _____.

    a. bronze   **b.** iron

    c. brass   **d.** gold

18.   The technique for silver brazing is essentially the same as for _____.

    a. brass   **b.** bronze

    c. iron   **d.** steel

19.   Chemical cleaning for aluminum can be done by dipping the pieces into a _____ bath.

    a. caustic   **b.** mild

    c. saltwater   **d.** water

20.   After the brazed joint has cooled, the _____ is (are) removed with a wire brush.

    a. slag   **b.** stains

    c. flux   **d.** contaminants

21.   The easiest way to bevel metal edges for braze welding is with a:

    a. hacksaw   **b.** file

    c. grinding wheel   **d.** hammer

22.   The placing of a thin coating of the rod metal on the parent metal is called:

    a. brazing   **b.** welding

    c. soldering   **d.** tinning

23.   What comes in handy to hold a joint steady during brazing?

    a. slip-joint pliers   **b.** a vise

    c. vise grips   **d.** none of the above

24.   Clean the finished joint after brazing and cooling with a _____.

    a. chemical compound   **b.** wire brush

    c. pile of ashes   **d.** long stick

25.   After the brazed pieces have cooled and are permanently bonded, clean with a:

    a. chemical bath to remove dirt and flux

    b. chemical bath to remove flux

    c. wire brush to remove slag

    d. none of the above

TRUE-FALSE EXAM

1. Brazing differs from other welding processes in that a nonferrous filler metal is used with a melting temperature of 840 degrees F.

    True   False

2. In brazing, since the base metals are not melted, the brazing metal is different in composition from the base metals.

    True   False

3. Heating a brazed joint above the melting point of the brazing metal, even if followed by subsequent cooling, may destroy the integrity of the joint.

    True   False

4. A brazed joint derives its strength from a combination of the braze metal, the base-metal alloy that is formed, and the penetration of the low-viscosity brazing metal into the grain boundaries of the base metal.

    True   False

5. Copper, when heated in a nitrogen atmosphere, requires virtually no clearance when being brazed.

    True   False

6. The most commonly used metals for brazing are copper, silver, and aluminum alloys.

    True   False

7. Copper is used only for brazing steel and other high-melting-point alloys such as high-speed steel and tungsten carbide.

    True   False

8. Spelter brass is another name for certain copper alloys.

    True   False

9. Silver alloys are widely used in fabricating copper and brass alloys.

    True   False

10. Silver alloys are used in brazing stainless steel.

    True   False

11.   A commonly used procedure in connection with brazing aluminum is to use sheets that have one or both surfaces coated with the brazing alloy to a thickness of about 10 percent of the total sheet thickness.

      True   False

12.   Fluxes play an important role in brazing by causing an increase in the torch temperature.

      True   False

13.   One of the primary factors affecting the quality and uniformity of brazed joints is cleanliness.

      True   False

14.   Although fluxes will dissolve modest amounts of oxides, they are not cleaners.

      True   False

15.   Borax is commonly used as a brazing flux.

      True   False

16.   Paste fluxes are usually utilized for furnace, induction, and dip brazing, while paste or powdered fluxes are used for torch brazing.

      True   False

17.   Although the base metal should be hot enough to melt the braze metal and assure that it remains molten and flows into the joint, the actual melting should be done with the torch.

      True   False

18.   Capillary action always will pull the molten braze metal into the smallest clearance, whether or not this was intended.

      True   False

19.   A common source of heat for brazing is a gas-flame torch.

      True   False

20.   Large amounts of brazing are done in controlled-atmosphere furnaces.

      True   False

21.   Furnace brazing is not used in mass production.

      True   False

22.   Furnace brazing is very economical, especially when no flux is used, as in brazing steel parts with copper.

      True   False

23. In salt-bath brazing the parts are heated by dipping them in a bath of molten salt water.

True   False

24. Dip brazing is used for small objects.

True   False

25. Induction brazing uses high-frequency currents for heating.

True   False

26. If maximum joint strength is desired, it is important to apply some pressure to the parts during heating.

True   False

27. Three types of brazed joints are butt, scarf, and shear.

True   False

28. Virtually all braze welding is done with an oxyacetylene torch.

True   False

29. Braze welding differs from brazing in that capillary attraction is used to distribute the filler metal.

True   False

30. Braze welding depends on gravity for filler-metal flow.

True   False

MULTIPLE-CHOICE ANSWERS

1.	D	6.	B	11.	D	16.	D	21.	C
2.	A	7.	B	12.	C	17.	A	22.	D
3.	A	8.	A	13.	A	18.	B	23.	C
4.	D	9.	B	14.	D	19.	A	24.	B
5.	C	10.	B	15.	D	20.	C	25.	C

TRUE-FALSE ANSWERS

1.	T	9.	F	17.	T	24.	T
2.	T	10.	T	18.	T	25.	T
3.	T	11.	T	19.	T	26.	T
4.	T	12.	F	20.	T	27.	T
5.	T	13.	T	21.	F	28.	T
6.	T	14.	T	22.	T	29.	F
7.	T	15.	T	23.	F	30.	T
8.	T	16.	T				

Chapter 12
SOLDERING

Soldering is a low-temperature joining process used to fasten two metal surfaces together without melting either of the metals. Fusion does not take place. A nonferrous filler metal, called solder or soft solder, is added to the joint where it melts at temperatures below 800 degrees F and flows into and along the joint surface by capillary action.

This joining process is also sometimes called soft soldering to distinguish it from brazing (sometimes referred to as hard soldering or silver soldering). Because of the lower temperatures there is less warpage of the metals than if they were welded.

APPLICATIONS

Soldering is a useful process for joining most metals in a broad range of applications. For example, it is widely used in the electronics industry, where its low temperatures reduce the possibility of damage to semiconductors and similar components. It is also widely used in HVAC or plumbing work to seal pipes and tubing.

Soldering processes include:

- Dip soldering
- Resistance soldering
- Wave soldering
- Infrared soldering
- Furnace soldering
- Iron soldering
- Torch soldering

Soldering advantages include the following:

- Self-contained and portable joining process
- Easy to learn
- Inexpensive equipment
- Applicable to most metals
- Low temperatures minimize possibility of damage to components being joined
- Rapid assembly of parts
- Mistakes are easily corrected
- Little or no finishing required

Soldering disadvantages include the following:

- Soldered joints are seldom used where strength is required.
- Solder and base metal colors may not match.

EQUIPMENT

Joints can be soldered by hand using either a gas torch or an electric soldering iron. Some commonly used types of soldering equipment are listed in Table 12.1.

TABLE 12.1 Manual soldering equipment.

Type	Comments
Gas torch	• The oxy-fuel gas torch uses oxygen in combination with natural gas or propane. • Acetylene is not recommended because the oxyacetylene flame is generally too hot for soft soldering. • The tip sizes used for soldering will vary depending on the thickness and type of metal being soldered. These tips will also frequently be designed for use with other types of surface work such as heating or silver brazing.
Soldering iron	• Consists of a large piece of copper, drawn to a point or edge and fastened to an iron rod having a handle grip. • Soldering irons are available in sizes ranging from the small lightweight types to the large heavy duty models found in industrial plants. • Soldering irons may be either internally or externally heated. External heating is usually done in ovens (furnaces) with a gas flame or other means. Internal methods of heating involve either gas or electricity, with the latter being the most common form.
Pencil iron	• Receives its name from its slim, pencil-like appearance. • Type most commonly used in non torch soldering. • Available in a variety of different sizes.
Hatchet iron	• Enables soldering solder in areas that are not easily accessible • Wide variety of tips are available. • Elements are interchangeable for a wide wattage range.
Soldering gun	• Quick-heating tool. • Used for repair work or intermittent soldering.
De-soldering tool	• Designed to melt and remove the solder by first heating it and then removing the molten solder by suction. • Vacuum bulb located above the heating iron sucks the molten solder from the surface. • Tips are heat-resistant and replaceable. • Tip orifices are available in a number of different sizes.
Iron holder	Designed to safely hold a soldering iron when not being used.

SOLDERS

Solders are nonferrous filler metals used to join metal surfaces. These soldering metals are formed from fusible alloys. There are a number of different types of solders, and the type used depends on the composition of the metals being joined.

Not all metals and metal alloys can be soldered. Among those that will not take a soldered joint are:

- Titanium
- Chromium
- High-tensile magnesium bronze
- Beryllium
- Cobalt
- Silicon

Classifying Solders

Some classifications of solders are based on the ASTM specification B32-58T. One category receives its designation from the silver content of the solder. For example, a 2.5S solder is one containing 97.5 percent lead and 2.3 to 2.7 percent silver. However, 2.5 percent is the desired level. These are the so-called lead-silver solders.

The largest classification category is designated according to its tin content (for example, a 5A solder represents one having 5 percent tin and 95 percent lead). There are three letter suffixes (A, B, and C) added to the numeral that indicates the tin percentage. These denote the following three composition classes:

- A: 0.12 percent maximum antimony content allowed for solders containing 35 percent tin or more
- B: 0.5 percent maximum antimony content allowed for solders containing 35 percent tin or more
- C: the antimony content may not exceed 6 percent of the tin content in the so-called tin-antimony solders (those containing 20 to 40 percent tin and an appropriate percentage of antimony)

By way of illustration, a 40A solder will consist of 40 percent tin, 60 percent lead, and a maximum of 0.12 percent antimony. The percentage given for tin is the desired percentage, and the one for lead is a nominal percentage. These amounts will vary to some degree. A 40B solder will contain 40 percent tin, 60 percent lead, and a maximum of 0.5 percent antimony. Finally, a 40C solder will contain 40 percent tin, 58 percent lead, and a maximum of 2.4 percent antimony (6 percent of the tin content).

Solder Types

Alloys are added to solder to produce filler metals with certain required characteristics:

- Antimony is added for extra strength (under specified conditions) and increased electrical conductivity.
- Silver is added to lead to improve wetting characteristics, particularly on copper or steel.
- Cadmium, silver, and zinc combinations produce a solder particularly suitable for joining aluminum or dissimilar metals.

Solders can be divided into a number of different groups according to their composition. The largest group and the one most commonly used contains the tin-lead alloy solders. **Note:** A solder containing too much lead makes a weak joint because the lead does not transfuse with brass. A solder containing too much tin becomes brittle.

Temperatures

The correct temperature for soldering must be determined for each particular job. This is done largely by experimenting and varies with the size of the work, the type of work, the solder composition, and the nature of the flux. Soldering ranges between 500 and 700 degrees F. Do not attempt to solder at temperatures near 800 degrees F. At approximately this temperature the drosses (the scum thrown off from molten metal) become soluble in the solder, and the solder is said to be "burnt" and will behave badly. Here are some specifics on working with solder temperatures:

- When soft solder is heated, it starts to melt at a specific temperature. The melting temperature will depend on the type of solder. As additional heat is applied, the temperature increases and more and more solder melts until it is all melted. The first temperature is called the solidus temperature (melting point), and the second temperature is called the liquidus temperature (flowing point). The latter temperature is important because the metal must be heated to a temperature higher than the flowing (liquidus) temperature or the solder will not flow.

- The ability of a solder to flow over a surface is one determination of the solderability of that metal. However, the flow characteristics will vary from one solder to the next. For example, tin-lead solder of, say, a 5 to 95 composition will flow quite a bit differently than a 95 to 5 tin-silver solder. Only practice will enable an operator to become accustomed to these differences.

- Soft-soldered joints cool and will solidify at their melting point. However, until the joint has cooled to a temperature below the melting point, no strain can be applied to the joint, since solder has no strength while it is still only partially solidified. Between the flowing and melting points is the so-called plastic range in which there is a mixture of solid and liquid solder.

- A few solders melt and flow at the same temperature. These are referred to as eutectic solders and are suitable for making bead joints but not for running into lap joints. It helps to know the melting and flowing points and the plastic range when changing from a solder of known composition to a substitute solder.

FLUXES

Metal surfaces are covered with thin, invisible oxides and other impurities. These must be removed before the solder is applied, or it will not adhere to the surface. They may be removed by applying a flux.

Note the following when working with fluxes:

- Do not skip the cleaning operation that precedes the application of the flux. Although the flux does remove oxides and other impurities, it is not a complete cleaning operation.

- A flux is a chemical that not only removes surface contaminants but also forms a thin film over the surface, preventing contact with the air. Moreover, the flux contributes to the free-flowing characteristic of the solder.

- No one flux can be assigned to any one metal as peculiarly suited to that metal for all purposes. The nature of the solder often determines the selection of the flux. In electrical work, for example, the best flux is made of pine-amber rosin because it does not cause corrosion.

- A corrosive flux, such as one containing zinc-chloride solution, should be strictly excluded from any electrical work.

- The Underwriter's Code permits the use of a flux composed of five parts zinc chloride, four parts alcohol, three parts glycerine, and water. This preparation permits the solder to flow freely and is not highly corrosive.

- Some flux manufacturers add wetting agents to their fluxes, having determined the amount to be added by experimental work.

- Wetting agents are penetrants that cause the solder film to thin out and cover a larger area. Fluxes with wetting agents are not necessary for 50-50 or 60-40 solders, but they are not detrimental. However, it is advisable to use fluxes with wetting agents for the lower tin solders. This is particularly necessary for long joints on copper or for all joints on brasses and bronzes.

Fluxes are commonly classified as either corrosive or noncorrosive. Some authorities recommended listing mild fluxes (slightly corrosive fluxes) as a third (intermediate) class of fluxes. However, because they result in some corrosion, they are often grouped with the corrosive fluxes.

SOLDERING-IRON METHOD

The soldering-iron or soldering-copper method is soldering done with a copper-tipped soldering iron. This is the oldest manual method of joining metals with solder.

Soldering-Iron Tips

The life expectancy of a copper-tipped soldering iron is commonly reduced by normal use, careless handling, and line voltage fluctuation. The three most common tip malfunctions that occur with normal use are tip erosion, tip corrosion, and tip freezing:

- Plain copper tips (i.e., those without special coatings) are most subject to tip erosion. During the soldering process, small bits of the tip break off and are carried away. This eventually destroys the effectiveness of the tip. Unfortunately, little can be done about this, and for this reason plain copper tips are not used for extensive production operations.

- Copper tips with coatings resistant to erosion are designed to prolong the life of the tip. This erosion-resistant coating is a thin plating of metal that can be tinned. It is also soluble in solder but to a much lesser extent than a plain copper tip. Oxides will form on the surface of a coated tip and must be removed. A wire brush is recommended for this purpose. Do not brush beyond the exposure of the original plating, or the tip will be ruined. The tip should be retinned immediately after brushing.

- The formation of oxides on the surface of the tip is known as tip corrosion. A wire brush or file is recommended for their removal. These oxides should be frequently removed; care should be taken not to remove the original coating; and the tip should be immediately retinned.

- Frequently while soldering scale will collect inside the core of the soldering iron. If the scale is not removed, it will solidify and freeze the tip, making it almost impossible to remove. The chance of this occurring can be reduced by removing the tip after soldering and cleaning off any scale that may have formed inside the core. A frozen tip will frequently result in the destruction of the heating element itself. Consequently, it is a wise move to prevent freezing from occurring in the first place.

Cleaning the Surface

Thoroughly cleaning the surface is, perhaps, the single most important step in soldering. The solder will not alloy to the surface of the base metal unless it is perfectly clean. Any surface contaminants (e.g., dirt, grease, etc.) will obstruct the alloying or wetting process of the solder.

Cleaning can be done mechanically or chemically. The former may be accomplished by machining, grinding, blasting (with sand, grit, or shot), or hand abrading. Generally, hand abrading is the method most commonly used in the small workshop and can be done in a number of ways, including rubbing with (fine-grade) steel wool, filing, brushing with a wire brush, or scraping.

Rub the metal surface until it has a bright, shiny appearance. The only exception to this rule is a tin surface or one on which a noncorrosive flux is to be applied. Both can be cleaned by wiping with a suitable solvent.

The surface can be chemically cleaned by wiping it with a clean cloth dipped in a solvent or detergent. This will remove most forms of grease, oil, or dirt. The chemical cleaning agent should be removed with hot water before applying the flux.

JOINT DESIGNS

A soldered joint does not have the strength of a welded joint, because the base metal does not melt and fuse with the filler metal. Consequently, a suitable joint design for soldering is determined by the stresses or loads to which it will be subjected. The operator should consider the degree to which the joint must withstand vibration, impact, tension, compression, and shear stress. The more pronounced these factors become, the greater is the need for such additional support as bolts, screws, or other fastening devices.

The electrical conductivity of a soldered joint is also an important consideration. Because so many soldered joints serve as electrical connections (e.g., on printed-circuit boards), one possessing poor electrical conductivity would hardly be acceptable. Both the composition of the solder and the finished joint must be conducive to a high degree of electrical conductivity. The operator must produce a soldered joint without points and avoid bridging electrical connections.

The lap and butt joints are the most common type of joint used for soldering. Of the two, the former is recommended and preferred by most operators. Figure 12.1 illustrates many of the typical joint designs used for soldering.

Note the following points about low-tin solders:

- Careful cleaning is particularly necessary with low-tin solders.

- Low-tin solders alloy more slowly, and so stronger flux solutions are necessary.

- Tight joints are particularly important with these solders.

- Low-tin solders require higher metal temperatures than the more commonly used 50-50 or 60-40 tin-lead solders.

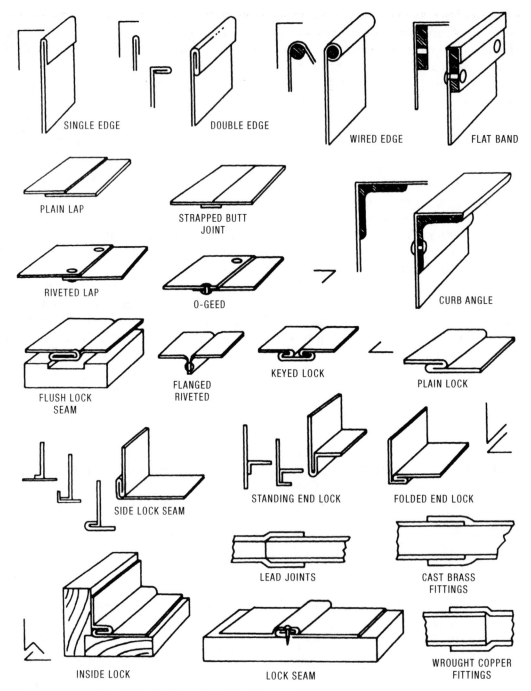

FIGURE 12.1 Recommended joint designs for soldering.

If the joint has a small enough diameter to be heated all over, such as in water-service tubing, there should be no trouble with low-tin solders if the foregoing procedure is used. However, on large-diameter joints, pretinning both sides of the joint after proper cleaning and fluxing is an additional help.

The same procedure is applied to high-lead silver solders except that a temperature about 250 degrees F higher than that for 50-50 solder is required to make the solder flow.

High-lead silver solders can be used where higher temperatures of application are not a handicap. However, because of the higher temperature required, the same flux that is used for 50-50 solder cannot be used for lead-silver solders. Special fluxes for this purpose are available, and flux manufacturers should be consulted.

Soldered Joints

Inspection of a soldered joint is generally done visually. Take a look at some of the soldered joints that fail to meet standards:

- Joints with excess solder

- Joints with no solder at all

- Joints with insufficient solder

- Flux residue in the joint

- A joint in which solder functions as a ground or causes a short

- A joint with points protruding from the solder

- Indications that the joint was made with a poor flow of solder

- Indications that the joint was moved before being completely cooled

Movement of the soldered joint before it has completely cooled causes it to be materially weakened. Premature movement often causes the solder to fracture. This in turn may result in joint failure.

Too much solder accounts for most rejections. Some operators apply entirely too much solder to the joint. This can be corrected before final inspection simply by removing the excess solder. Dipping the scrap end of an unsoldered standard conductor into flux and pressing it between the terminal and the hot iron tip will cause the excess solder to be drawn off.

Insufficient solder on a joint can cause a weak joint. These joints, along with those to which no solder has been applied, are regarded as not being repairable and are rejected.

Excess flux sometimes remains in the joint after soldering. If the soldered parts, such as wire, tube, or sheet, are loose, the joint is defective and must be rejected. However, if no movement is possible, the flux can be removed by reheating.

Torch Soldering

Oxyacetylene torches are generally not recommended for soldering. That is because the temperature of the flame is too high. It usually results in destroying the protective action of the flux. However, this can be avoided by directing the full force of the flame away from the spot being soldered. This form of indirect heating can also be used with other types of gas-flame soldering. Lower flame temperatures can be achieved by using:

- Propane

- Natural gas

- Butane

Oxygen burned with a fuel gas will produce higher temperatures than air burned with the same gases. The former will also produce a sharper, more well-defined flame than a torch operating on the Bunsen-burner principle. The latter are generally gasoline-filled torches, and they produce a widespread flame that is somewhat bushy in appearance.

MULTIPLE-CHOICE EXAM

1. There are three ways of joining metal: soldering, _____, and welding.

    a. epoxying   **b.** gluing

    c. brazing   **d.** none of the above

2. The strongest of the three methods is:

    a. epoxying   **b.** welding

    c. brazing   **d.** soldering

3. The weakest of the three methods is:

    a. epoxying   **b.** welding

    c. brazing   **d.** soldering

4. Soldering is similar to _____ in that a nonferrous metal is used to bridge the space between the two metals.

    a. brazing   **b.** welding

    c. gluing   **d.** none of the above

5. What is the simplest type of soldering tool?

    a. tinner   **b.** torch

    c. fluxer   **d.** soldering iron

6. Old soldering irons had no power of their own. They had to be heated in a special _____ that was most commonly gas-fired.

    a. solder microwave   **b.** furnace

    c. brazer   **d.** torch

7. The propane _____ is the tool of choice for soldering the joints in a copper-pipe plumbing system.

    a. torch   **b.** tip

    c. bottle   **d.** flame

8. The hand-held soldering gun is nothing more than a short-circuited _____.

    a. transformer   **b.** transistor

    c. tip   **d.** transducer

9. Another gas that is sometimes used in soldering is:

   a. acetylene **b.** natural

   c. MAPP **d.** propane

10. Radiator repair shops use a _____ for soldering the tank to the core.

    a. soldering torch **b.** soldering gun

    c. soldering iron **d.** soldering copper

11. One indication that the soldering-iron tip is overheating is a _____ tinge that will appear when the tip is hotter than it should be.

    a. bluish **b.** greenish

    c. reddish **d.** brownish

12. The soldering iron produces better results if the tip is coated with a thin layer of _____.

    a. tin **b.** copper oxide

    c. solder **d.** magnesium

13. What solder is used for car radiators?

    a. 60A **b.** 30A

    c. 20A **d.** 50A

14. Recently an alternative to lead solder was developed. It is called:

    a. zinc solder **b.** tin solder

    c. leadless solder **d.** none of the above

15. Fluxes are available in powder, paste, and _____ form.

    a. liquid **b.** fluid

    c. foam **d.** solid

16. All fluxes are either corrosive or _____.

    a. free-flowing **b.** noncorrosive

    c. thick and hard **d.** soft and bendable

17. Which type of ingredients do corrosive fluxes have?

    a. inactive   **b.** active

    c. organic   **d.** inorganic

18. Corrosive fluxes are not suited for _____ work.

    a. copper   **b.** electronic

    c. plumbing   **d.** plastic

19. Probably the most common type of corrosive flux is called:

    a. acid flux   **b.** rosin-core flux

    c. hard-core flux   **d.** none of the above

20. The flux usually found in the core of 50-50 solder is:

    a. corrosive   **b.** noncorrosive

    c. fluxless   **d.** rosin

21. Which solder is used for electronic circuit boards?

    a. 60-40 rosin core   **b.** 50-50 acid core

    c. 40-60 rosin core   **d.** 60-40 acid core

22. The first rule of successful soldering is _____.

    a. use acid-core solder   **b.** cleanliness

    c. flux   **d.** proper heat

23. In soldering it is best to avoid the _____ joint.

    a. plain-lap   **b.** fillet

    c. lap   **d.** butt

24. Clean the tip of a soldering iron with steel wool before tinning it with _____.

    a. solder   **b.** brazing rod

    c. tin   **d.** zinc

25. Remove corrosive flux from a cooled solder joint with a _____ cloth.

    a. damp   **b.** dry

    c. soaked   **d.** none of the above

26. The key to successful soldering lies in choosing the proper _____ for the project at hand.

    a. torch   **b.** tip

    c. flux   **d.** solder

27. Which is the solder that has greatest strength quality because of its high lead content?

    a. 50–50   **b.** 60–40

    c. 40–60   **d.** 5A

TRUE-FALSE EXAM

1. Prolonged inhalation of corrosive-flux fumes should be avoided.

True   False

2. Before the solder hardens, wipe the joint with a damp cloth.

True   False

3. Solder is also known as a fusible alloy.

True   False

4. Most solders are alloys of lead and tin with a very small amount of copper.

True   False

5. For higher-temperature service, lead-silver solders are used.

True   False

6. Soldering fluxes are classified as corrosive and more corrosive.

True   False

7. The two most commonly used corrosive-type fluxes are muriatic acid and a mixture of zinc and ammonium chlorides.

True   False

8. Dip soldering is used exclusively for wire ends.

True   False

9.   Soldered joints seldom will develop strength in excess of 1.72 MPa (250 psi).

         True   False

10.   Butt joints should never be used in soldering.

         True   False

11.   When soldering copper water lines, it is important not to use too much solder.

         True   False

12.   Probably the most common use of the torch in soldering is in copper plumbing joints.

         True   False

13.   There is no need to remove corrosive fluxes after the joint has cooled.

         True   False

14.   The most common type of corrosive flux is called acid flux.

         True   False

15.   Acid-core solder is used for plumbing and for like metals.

         True   False

16.   Rosin-core solder is used for electronics.

         True   False

17.   Fluxes can be grouped into corrosive and noncorrosive categories.

         True   False

18.   Because of its composition, leadless solder behaves slightly differently from traditional lead solder.

         True   False

19.   When using the soldering iron, it is best to keep the tip coated or tinned.

         True   False

20.   Most soldering requires a temperature over 800 degrees Fahrenheit.

         True   False

21.   The solder known as 5A is good for torch soldering.

         True   False

22. The soldering gun is quick to heat to melting temperature, inasmuch as it is nothing more than a copper strip used to short-circuit a transformer.

True   False

23. MAPP gas is another bottled gas that can be used to provide torches with a source of energy.

True   False

24. Propane is the most common source of energy for the soldering torch used by the do-it-yourselfer.

True   False

25. Plumbers have used propane torches for soldering pipes for years.

True   False

MULTIPLE-CHOICE ANSWERS

1. C	7. A	13. C	18. B	23. D
2. B	8. A	14. C	19. A	24. A
3. D	9. C	15. A	20. A	25. A
4. A	10. A	16. B	21. A	26. D
5. D	11. B	17. D	22. B	27. D
6. B	12. C			

TRUE-FALSE ANSWERS

1. T	8. T	14. T	20. F
2. T	9. T	15. T	21. T
3. T	10. T	16. T	22. T
4. F	11. T	17. T	23. T
5. T	12. T	18. T	24. T
6. F	13. F	19. T	25. T
7. T			

Chapter 13
LEAD, TIN, AND ZINC

L
ead, tin, and zinc are soft metals. They have a strong resistance to corrosion. However, they have low strength and melting temperatures well below 1000 degrees F.

LEAD AND LEAD ALLOYS

Lead is described as a grayish-white metal. It is noted for its softness, ductility, and resistance to corrosion. Lead is also characterized by low strength, which includes low fatigue strength. It also has a tendency to fatigue cracking. Lead can be obtained with a purity of 99 percent. The melting point of pure lead is 621 degrees F (327.5degrees C).

Lead has many applications. Pipes and other fittings account for about 60 percent of the annual production and are used in:

- Plumbing
- Shielding protection against x-ray and gamma radiation in the dental and medical occupations
- Corrosion-resistant linings
- Cable sheathings
- As an alloy with tin in some low-melting types of solder
- Storage batteries

Lead is commonly welded with the oxyacetylene process. It can also be joined by soldering.

Welding and Joining Tips

Following is a handy outline for using oxyacetylene welding and soldering with lead:

1. Oxyacetylene Welding
   a. Welding can be performed in all positions.
   b. The welds are normally stronger than the base metal.
   c. No flux is required, and edge preparation is relatively easy because the metal is very soft.
   d. Use a small torch with small tip sizes.
   e. Use a neutral flame with a slight acetylene excess.
   f. Use a gas pressure less than 5 psi for best results. **Note:** The gas pressure will depend ultimately on the size and type of weld.
   g. Use a filler rod (when required) matching the analysis of the base metal. A filler rod is not necessary when making flange and edge joints.
   h. Pull back the torch when the weld pool becomes too molten. Doing so will avoid excess fusion, excess fluidity, and burn through, all of which are problems directly related to the low melting point of lead.
   i. Other gas-welding processes used to weld lead include oxyhydrogen welding and air-acetylene welding. Natural gas in combination with oxygen is also used.
   j. The oxyhydrogen welding process is best suited for thinner pieces of lead (generally less than 1/4 inch thick). Thicker pieces of lead are commonly welded with an oxyacetylene torch because of its hotter flame.

2. Soldering

    a. Never use solder to join containers, piping, or tubing used to store, carry, or transport corrosive chemicals. Materials for these uses must be welded.

    b. Joint fit-up is critical. Maintain close joint tolerances and clamp the sheets or parts in place.

    c. Thoroughly clean the surface before soldering. A clean surface is necessary to promote capillary attraction - the motive force that draws the solder into the narrow joint.

    d. Rosin fluxes are recommended for soldering lead.

    e. No postsoldering treatment is required.

TIN AND TIN ALLOYS

Tin is silvery white in color. It has a melting point of approximately 449 degrees F (232 degrees C). Tin is known for its softness and ductility.

Tin is combined with many other metals to form commonly used metal alloys. For example, tin is mixed with copper to form bronze and with lead to form solder. Tinfoil is produced from a rolled lead-tin alloy.

The tin cans used in the food industry consist of thin iron sheets covered on both sides by a coating of tin. Tin is used because it is a nontoxic metal. Pewter is an alloy consisting of at least 91 percent tin and small amounts of antimony and copper. This is only a small portion of the many metals in which tin functions as an alloying element. Some of the others include (in percentages):

- Aluminum-tin alloy (79.0 aluminum, 20.0 tin, 1.0 copper), used in the automotive industry for the production of such items as camshaft bearings and connecting rods

- Phosphor bronze (95.0 copper, 5.0 tin)

- Gun metal (88.0 copper, 10.0 tin, 2.0 zinc), used for the production of gears and bearings

Note the following tips about tin:

- Tin alloys are welded with the same techniques used to weld lead alloys.

- The reaction to heat and other characteristics of the welding process are also similar.

- Pure tin is rarely welded.

ZINC AND ZINC ALLOYS

Zinc is obtained from a sulfide ore by the electrolytic process or by other methods. Pure zinc has a melting point of 787 degrees F (421 degrees C). Its tensile strength is approximately 18,000 psi, but the tensile strength of cast zinc is considerably higher at 40,300 psi.

Zinc is used in its pure form for galvanizing. It is also combined with other metals (aluminum, magnesium, iron, and tin) to form zinc castings. In addition, zinc serves as an alloying element in copper-zinc alloys such as zinc bronze.

Note: Zinc fumes have a nauseating effect and can pose a significant health hazard. These fumes are released in the form of a white vapor at certain temperatures. Several methods can be used for reducing zinc fumes, including the use of an oxidizing flame (when welding with oxyacetylene), special fluxes, and welding at lower temperatures.

Zinc as a pure metal has one important use: galvanizing iron and steel. Galvanizing accounts for 35 percent of all zinc used. Sherardizing is similar to galvanizing, but the coating process is somewhat different. The process utilized puts a diffusion coating on iron and steel.

Zinc is also used in die casting, which accounts for over 40 percent of the annual consumption of zinc.

Zinc has another use in the commercial world. It is part of the cooper-zinc alloy that is used to make brass and bronze. This accounts for about 15 percent of the annual production of zinc.

MULTIPLE-CHOICE EXAM

1. Lead, tin, and zinc are known as _____ metals.
    a. rough   b. tough
    c. hard   d. soft

2. Lead is characterized by low strength with a low fatigue _____.
    a. level   b. point
    c. strength   d. peeling

3. Where is lead used?
    a. plumbing   b. storage batteries
    c. cable sheathings   d. all of the above
    e. none of the above

4. Zinc has a very useful life in alloying with copper to make _____.
    a. steel   b. bronze
    c. brass   d. manganese

5. Lead is commonly welded with the _____- welding process.
    a. arc   b. oxyacetylene
    c. electric   d. chemical

6. For best results in torch welding of lead, use a gas pressure less than _____ psi.
    a. 15   b. 5
    c. 10   d. 20

7. No flux is used in lead welding because the metal is very _____.

    a. soft
    b. hard

    c. smooth
    d. rough

8. The oxyhydrogen process is best suited for welding the _____ lead pieces.

    a. thinner
    b. thicker

    c. foil
    d. none of the above

9. Never use solder to join containers, piping, or tubing used to store, carry, or transport _____ chemicals.

    a. noncorrosive
    b. corrosive

    c. active
    d. inactive

10. Tin is used in the food industry because it is _____.

    a. a colorful metal
    b. a toxic metal

    c. a nontoxic metal
    d. none of the above

11. Tin is _____ in color.

    a. silvery white
    b. silvery gray

    c. gray
    d. silver

12. Tin alloys are welded with the same techniques as used to weld _____ alloys.

    a. lead
    b. magnesium

    c. steel
    d. brass

13. Pure tin is rarely _____.

    a. brazed
    b. soldered

    c. welded
    d. none of the above

14. Zinc is used in its pure form for _____.

    a. galvanizing
    b. jewelry

    c. car trim
    d. none of the above

15. Zinc is used in _____ casting.

    a. ceramic
    b. die

    c. fish
    d. none of the above

TRUE-FALSE EXAM

1.   Lead, tin, and zinc are soft metals that have low strength and low melting temperatures.
       True   False

2.   Lead is commonly welded with the oxyacetylene-welding process.
       True   False

3.   Welding of lead can be done in all positions.
       True   False

4.   A filler rod is not necessary when making flange and edge joints with lead.
       True   False

5.   Lead has a melting temperature of 621 degrees F.
       True   False

6.   Tin is a gray-colored metal and very hard.
       True   False

7.   Tin is mixed with copper to make bronze.
       True   False

8.   Tinfoil is produced from a rolled lead-tin alloy.
       True   False

9.   Phosphor bronze is 85 percent copper and 15 percent tin.
       True   False

10.   Gun metal (88 percent copper, 10 percent tin, and 2 percent zinc) is used for the production of gears and bearings.
       True   False

11.   Tin alloys are welded with the same techniques used to weld lead alloys.
       True   False

12. Pure tin is often welded.

True   False

13. Zinc is obtained from a sulfide ore by the electrolytic process.

True   False

14. Zinc fumes have a nauseating effect and can pose a health hazard.

True   False

15. About 50 percent of annual consumption of zinc goes for die-casting projects.

True   False

MULTIPLE-CHOICE ANSWERS

1.	D	4.	C	7.	A	10.	C	13.	C
2.	C	5.	B	8.	A	11.	A	14.	A
3.	D	6.	B	9.	B	12.	A	15.	B

TRUE-FALSE ANSWERS

1.	T	5.	T	9.	F	13.	T
2.	T	6.	F	10.	T	14.	T
3.	T	7.	T	11.	T	15.	F
4.	T	8.	T	12.	T		

Chapter 14

IDENTIFYING METALS

Welding Licensing Exam Study Guide

Correctly identifying the base metal provides the welder with the necessary information to choose an appropriate welding process. It also allows the welder to:

- Decide on the type of weld to use
- Determine the type of weld preparation
- Choose the composition of the filler rod (if used)

The inability to correctly identify the metal greatly increases the possibility of a poor weld joint.

Sending a metal to a laboratory for chemical analysis and/or machine testing is expensive and time-consuming. This makes it highly impractical for small welding shops and the individual welder.

The metal-identification methods described here are not comprehensive, but they can be useful in narrowing down the alternatives. Ferrous metals, for example, can be identified by the spark test. Nonferrous metals do not produce spark streams.

IDENTIFY A METAL BY USE

Metals are often easily identified by how they are used. Engine blocks, gears, shafts, and the like can be expected to be constructed from certain types of metal (e.g., cast iron, forged steel, or in some cases aluminum engine blocks). The metal may be specified by the manufacturer in the literature accompanying the component or part, or such information may be obtained by contacting to the manufacturer.

IDENTIFY BY SPECIAL MARKS

Codes or symbols are stamped on some metals by the supplier. For instance, nickel alloys, such as Monel and Inconel, and wrought iron are easily identified. The metal may also be identified in the drawing requirements (bill of materials) for a particular application.

Where possible, try to obtain a certificate for identification of the metal used in a particular project. Material certificates should include the manufacturer's name, commercial brand name, heat number, dimensions, chemical analysis (composition), mechanical properties, delivery hardness, and finish.

SURFACE-CONDITION IDENTIFICATION

It is sometimes possible to identify a metal by how the surface feels to the touch. For example, copper, brass, bronze, nickel, Monel, Inconel, lead, and wrought iron are very smooth.

On the other hand, stainless steel is slightly rough to the touch, and both low-alloy and high-carbon steels often have forging marks on the surface that make them seem a bit rough.

FINISHED AND UNFINISHED COLOR IDENTIFICATION

The finished (machined) and unfinished color of a metal can be used as a means of identification. Table 14.1 indicates color identifications for a number of commonly used metals and metal alloys. These colors are of unfinished, nonfractured metal surfaces and finished (machined) surfaces. This type of identification is not as accurate as some others, especially with the various steel types, Monel metal, and nickel.

TABLE 14.1   Finished and unfinished surfaces of selected metals by color.

Metal type	Unfinished nonfractured surface color	Finished (machined) surface color
Alloy steel	Dark gray	Bright gray
Aluminum and aluminum alloys	Very light gray	Very white
Beryllium	Steel gray	—
Brass	Green, brown, or yellow	Red to whitish yellow
Bronze	Green, brown, or yellow	Red to whitish yellow
Cast steel	Dark gray	Bright gray
Chromium	Steel gray	—
Columbium	Yellowish-white	—
Copper	Reddish brown to green	Bright copper red
Gray cast iron	Dark gray	Light gray
High carbon steel	Dark gray	Bright gray
Inconel	White	—
Lead	White to gray	White
Low alloy steel	Blue gray	Bright gray
Low carbon steel	Dark gray	Bright gray
Magnesium	Silver white	—
Malleable iron	Dull gray	Light gray
Manganese steel (under 2% manganese)	Gray white (resembles color of iron)	—
Manganese steel (14%)	Dull	—
Medium carbon steel	Dark gray	Bright gray
Molybdenum	Silvery white	—
Monel metal	Dark gray, smooth	Light gray
Nickel	Dark gray	White
Tantalum	Gray	—
Tin	Silver white	—
Titanium	Steel gray	—
Tungsten	Steel gray	—
Vanadium	Bright white	—
White cast iron	Dull gray	Seldom machined
Wrought iron	Light gray	Light gray
Zinc	Dark grey	—
Zirconium	Silver white	—

FIELD-TEST IDENTIFICATION

Welders in the field or in small shops must rely on other methods to identify the metal. This means they will be required to rely on their own experience plus some useful field tests. However, these tests are not as accurate as chemical analysis or machine testing. Nevertheless, they are very useful. These tests include:

1. Fracture test
2. Spark test
3. Flame test
4. Chip test
5. Magnetic test
6. File-scratch test

Fracture Test

A chisel or hammer can be used to remove a chip from the metal to be welded. Identification can then be made by the color of the fractured surface. Table 14.2 can serve as a guide in identification.

TABLE 14.2 Fracture test.

Metal type	Comments
Aluminum and aluminum alloys	White
Brass	Not used
Bronze	Not used
Cast steel	Bright gray
Copper alloys	Not used
Deoxidized copper	Red
Gray cast iron	Brittle fracture; dark gray fractured surface
High carbon steel	Very Light gray
High manganese steel (11-14% manganese)	Very coarse grained
Inconel	Light gray
Lead	White, crystal grain structure
Low alloy steel	Medium gray
Low carbon steel	Bright gray
Magnesium	Not used
Malleable cast iron	Brittle fracture surface; dark gray
Manganese steel (14%)	Coarse grained fracture
Medium carbon steel	Very light gray
Monel	Light gray
Nickel	Almost white to white
Nodular cast iron	Brittle fracture surface
Stainless steel (300 series)	Varies widely depending on type
Stainless steel (400 series)	Varies widely depending on type
Tin	Not used
Titanium	Not used
Tungsten	Brittle grain
White cast iron	Brittle
Wrought iron	Bright gray, fibrous grain pattern

Spark Test

Ferrous metals can be identified by the spark test. This is neither a difficult nor an expensive method of testing, but it does require knowledge of different spark stream characteristics.

Lightly touching the surface of the metal to a spinning power-driven grinding (emery) wheel produces a spark stream (pattern). This is then observed against a black background. Spark streams will vary in color, volume, length, and size depending on the carbon content of the metal. (See Figure 14.1 and Table 14.3.)

Note: The length of the spark stream results more from the amount of pressure exerted against the grinding wheel and is not a property of the metal.

Caution: A spark test may produce an electrical spark that can cause an explosion.

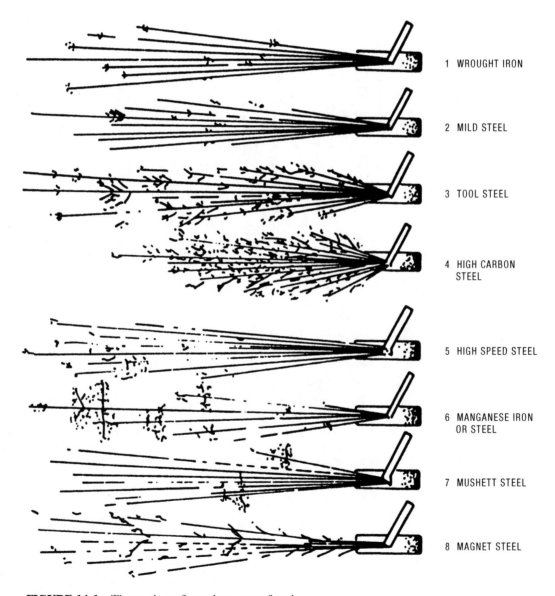

1 WROUGHT IRON

2 MILD STEEL

3 TOOL STEEL

4 HIGH CARBON STEEL

5 HIGH SPEED STEEL

6 MANGANESE IRON OR STEEL

7 MUSHETT STEEL

8 MAGNET STEEL

FIGURE 14.1 The spark test for various types of steels.

TABLE 14.3 Spark test for selected metals.

Metal type	Comments
Alloy steels	Spark stream similar to plain carbon steel except for changes proportional to the amount of alloy or alloys. Color changes from straw to deep orange or to red. The stream pattern length shortens with carrier lines appearing heavier. Carbon spark bursts reduced in number.
Aluminum and aluminum alloys	Not used
Brass	Not used
Bronze	Not used
Chromium	The particles seem to follow a broken line with a very slight explosion. Just before they disappear, the color is of chrome yellow and shows no trace of a carbon spark.
Copper	Not used
Copper alloys	Not used
Gray cast iron	Small volume of stream. Red carrier lines with sprigs or spark bursts similar to low carbon steel.
High carbon steel (Carbon content of .45% or more)	Yellow carrier lines with numerous bright spark bursts. The iron lines of the spark stream are practically eliminated, with an increase of the star-like explosions which often divide and subdivide, causing a beautiful display of sparks.
High manganese steels (11-14% manganese)	Not used
High sulphur steel	Carrier lines swell. Sprigs and spark bursts not very similar and similar to low carbon steel.
Inconel	Not used
Lead	Not used
Low alloy steel	The spark stream is characterized by a division or forking of the luminous streak.
Low carbon steel (Carbon content of .20% or less).	Long yellow carrier lines with little noticeable spark bursts.
Malleable cast iron (annealed)	Moderate stream volume. Stream color red changing to yellow at end. Many fine, repeating spurts.
Manganese steels (under 2% manganese)	Large bright white spark bursts spread out in a fan shape.
Manganese tool steel	Differs widely from the carbon spark by shooting or exploding at right angles from its line of force. Each dart of the manganese spark is subdivided into a number of white globules.
Medium carbon steel (.20% to .45% carbon)	Noticeable sprigs or spark bursts forming at ends of yellow lines.
Monel	Not used
Nickel	Very small stream volume. Stream orange in color. No sputs.
Nodular cast iron	Spark stream has lighter to orange section near the grinding wheel. Lower section is longer than gray cast iron, fewer spark bursts, and a near yellow color.
Stainless steel (300 series)	Various depending on alloy content
Stainless steel (400 series)	Various depending on alloy content
Tin	Not used
Tool steel (lower grades: 50–100 point carbon content)	The spark stream is characterized by iron lines that become less and less conspicuous. The forking of the luminous streak occurs very much more frequently, often subdividing. The lower the carbon content, the fewer the sparks and the further these sparks occur from the source of heat.
Tungsten	Not used
Tungsten-chromium die steel	Small stream volume. Red stream with many, fine repeating spurts.
White cast iron	Very small stream volume. Stream color red changing to yellow at end. Many fine, repeating spurts.
Wrought iron	Wrought iron creates a spark stream consisting of small particles flowing away from the point of contact in a straight line. The stream becomes broader and more luminous some distance from the source of heat and then the particles disappear as they started.
Zinc	Not used

Flame Test

Metals turn a distinctive color when a torch flame is directed against the surface. (See Table 14.4.) Metals also melt at different rates. Their molten weld puddles and slag may differ.

Note: Do not use the flame test if you suspect that the metal is magnesium. Magnesium will ignite under a torch flame in the open atmosphere.

TABLE 14.4   Flame test for selected metals.

Metal type	Comments
Alloy steel	Impossible to establish uniform flame characteristics because of the widely varying composition of the alloy steels.
Aluminum and aluminum alloys	Melts without any color.
Brass	Not used
Bronze	Not used
Cast iron	Melts slowly
Copper	Not used
Copper alloys	Not used
Gray cast iron	
High carbon steel	Turns a bright red just before melting and then melts quickly.
High manganese steels (11-14% manganese)	Shows color
High sulphur steel	Turns a bright red just before melting and then melts quickly.
Iconel	Not used
Lead	Melts fast
Low alloy steel	Turns a bright red just before melting and then melts quickly.
Low carbon steel	Turns a bright red just before melting and then melts quickly.
Magnesium	Ignites when it comes in contact with air and a flame.
Medium carbon steel	Turns a bright red just before melting and then melts quickly.
Monel	Not used
Nickel	Not used
Nickel alloys	Not used
Stainless steel (300 series)	Melts fast
Stainless steel (400 series)	Melts fast
Tin	Melts fast
Titanium	Not used
White cast ion	Turns a dull red before melting and then melts slowly.
Wrought iron	Turns a bright red just before melting and then melts quickly.
Zinc	Melts fast

Chip Test

Use a cold chisel and hammer to cut the metal for the chip test. The metal is identified by how easy or difficult it is to cut the surface, whether the cut edges are smooth or rough, and whether the cut is continuous or results in brittle, small chips. (See Table 14.5.)

TABLE 14.5   Chip test for selected metals.

Metal type	Comments
Alloy steel	Impossible to establish uniform chip characteristics because of the widely varying composition of the alloy steels.
Aluminum and aluminum alloys	Chips easily. Very smooth chips. saw edges where cut. Chip can be as long and continuous as desired.
Brass	Chips easily. Very smooth chips. saw edges where cut. Chip can be as long and continuous as desired.
Bronze	Chips easily. Very smooth chips. saw edges where cut. Chip can be as long and continuous as desired.
Copper	Chips easily. Very smooth chips. saw edges where cut. Chip size can be as long and continuous as desired.
Gray cast iron	Difficult to chip. Chips break off from base metal in small fragments.
High carbon steel	Very hard metal and difficult to chip. Chips can be any size or shape, continuous and as long as desired. Edges are lighter in color than low carbon steels.
High sulphur steel	Chips easily. Chisel cuts a continuous chip with smooth edges.
Lead	Chips very easily. Chips can be any size or shape and as long as desired.
Low alloy steel	
Low carbon steel	Chips easily with smooth edges. Chisel can cut a continuous chip if desired.
Malleable cast iron	Tougher metal than other cast irons. Difficult to chip. Chips do not break short.
Medium carbon steel	Chips easily. Chisel cuts a continuous chip with smooth edges.
Molybdenum	
Monel metal	Chips easily. Chips have smooth edges and can be as long as desired.
Nickel	Chips easily. Chips have smooth edges and can be as long as desired.
Nickel alloys	Chips easily. Chips have smooth edges and can be as long as desired.
White cast ion	Chipping results in small broken fragments. Brittleness of metal prevents formation of chip path with smooth edges.
Wrought iron	Chips easily. Very smooth edges on cut side of chip. Chip size can be as long and continuous as desired.

Magnetic Test

Use a small magnet against the surface of the metal (Table 14.6). If the magnet fails to attach itself to the surface, the metal is clearly nonmagnetic. Nonmagnetic metals are the nonferrous types, including copper and copper alloys, aluminum and aluminum alloys, and magnesium. Ferrous metals, such as carbon steels, low-alloy steels, pure nickel, and martensitic stainless steels, are strongly magnetic. The magnet will attach itself securely to the metal surface. Some metals, however, are only slightly magnetic, but experience will eventually enable you to distinguish between strongly and slightly magnetic metals.

TABLE 14.6   Magnetic test for selected metals.

Metal type	Comments
Alloy steel	Reduced magnetism depending on the alloying element.
Aluminum and aluminum alloys	Nonmagnetic
Brass	Nonmagnetic
Bronze	Nonmagnetic
Cast steel	Magnetic
Copper	Nonmagnetic
Gray cast iron	Strongly magnetic
High carbon steel	Strongly magnetic
High sulphur steel	Strongly magnetic
Iconel	Nonmagnetic
Lead	Nonmagnetic
Low carbon steel	Strongly magnetic
Magnesium	Nonmagnetic
Malleable cast iron	Strongly magnetic
Manganese steels (under 2% manganese)	Magnetic
Medium carbon steel	Strongly magnetic
Monel metal	Slightly magnetic
Nickel	Magnetic
Nickel alloys	Some are magnetic, some are weakly magnetic, others are nonmagnetic. Their magnetism can seldom be determined from analysis.
Stainless steel (300 series)	Nonmagnetic
Stainless steel (400 series)	Magnetic
Tin	Nonmagnetic
Titanium	Nonmagnetic
White cast iron	Strongly magnetic
Wrought iron	Strongly magnetic
Zinc	Nonmagnetic

File Scratch Test

A simple scratch test with a sharp metal mill file is sometimes used to make a rough determination of metal hardness to identify different types of steels. The file is drawn across the metal surface, and the scratch is compared to selected Brinell values to determine the hardness of the metal. (See Table 14.7.)

TABLE 14.7   Brinell values compared with a file scratch test.

Brinell hardness test	File action	Probable metal type
100 BHN	File bites into surface very easily	Low carbon steel
200 BHN	File removes metal with slightly more pressure	Medium carbon steel
300 BHN	Metal exhibits its first real resistance to the rile	High carbon steel or high alloy steel
400 BHN	File removes metal with difficulty	Unhardened tool steel
500 BHN	File marks the surface and just barely removes metal	Hardened tool steel
600 BHN	File slides over surface without removing metal; file teeth are dulled	Metal is harder than the file

Courtesy James F. Lincoln Arc Welding Foundation

MULTIPLE-CHOICE EXAM

1. What is the first step in welding?

   a. getting the torch lit

   b. getting the proper tip for the job

   c. identifying the metal to be welded

   d. none of the above

2. The easiest and most reliable way to identify metal is to:

   a. use the color test

   b. use the past experience method

   c. use the spark test

   d. send it to a lab for testing

3. Which metal type can you recognize by touch or feel?

   a. copper, brass, bronze, nickel, Monel, Iconel, lead and wrought iron—they are very smooth

   b. all metals are very rough to the touch

   c. all metals are slightly rough to the touch

   d. none of the above

4. Which of these metals have a light gray color when machined?

   a. tin, high-carbon steel, vanadium

   b. cast iron, malleable iron, Monel metal, wrought iron

   c. chromium, copper, aluminum

   d. cast steel, alloy steel, medium-carbon steel

5. Which of the following is not a field test for identifying metal:

   a. spark test   **b.** drop test

   c. flame test   **d.** chip test

   e. magnetic test   **f.** file scratch test

6. The fracture test is made by studying a:

   a. piece of magnesium   **b.** spark stream

   c. chip of the metal   **d.** none of the above

7.   Which of the field tests can be very unsafe?

    a. magnetic test   **b.** fracture test

    c. spark test   **d.** none of the above

8.   Which of the following metals cannot be tested by the spark test?

    a. Iconel   **b.** lead

    c. bronze   **d.** all of the above

9.   Which of the following metals cannot be tested by the flame test?

    a. brass   **b.** copper

    c. bronze   **d.** all of the above

10.   How is the chip for the chip test obtained?

    a. by using a torch and chisel

    b. by using a hammer and chisel

    c. by using a mallet and file

    d. by using a file and vise

11.   The magnetic test is made by using a small magnet to see if it will stick to the metal. Which of these metals will test magnetic?

    a. malleable cast iron   **b.** stainless steel (400 series)

    c. white cast iron   **d.** all of the above

    e. none of the above

12.   A simple scratch test with a sharp metal mill file is sometimes used to make a rough determination of metal hardness to identify different types of _____.

    a. iron   **b.** steel

    c. copper   **d.** none of the above

TRUE-FALSE EXAM

1. The first step in welding is to identify the metal being welded.

    True   False

2. Ferrous metals can be identified by the spark test.

    True   False

3. Nonferrous metals can be identified by the spark test.

    True   False

4. Sometimes it is possible to identify a metal by how the surface feels to the touch.

    True   False

5. The color of an unfinished nonfractured alloy steel is bright orange.

    True   False

6. Malleable iron is silvery dull gray.

    True   False

7. The spark test is very useful in identifying magnesium.

    True   False

8. When cast steel is broken and a chip is examined, it will probably be bright gray.

    True   False

9. The spark test can be done with a grinding wheel or a steel file.

    True   False

10. Manganese metal will easily start to combust in the presents of sparks.

    True   False

11. A simple scratch test with a sharp metal mill file is sometimes used to make a rough determination of hardness of different types of steels.

    True   False

12. When the metal is harder than the file, the file will slide over the surface of the piece of steel without removing any metal and the file teeth will become dull; this is an indication that the steel is probably reading 600 BHN on the Brinell hardness test.

    True   False

MULTIPLE-CHOICE ANSWERS

1.	C	4.	B	7.	C	9.	D	11.	D
2.	D	5.	B	8.	D	10.	B	12.	B
3.	A	6.	C						

TRUE-FALSE ANSWERS

1.	T	4.	T	7.	F	10.	F
2.	T	5.	F	8.	T	11.	T
3.	F	6.	F	9.	F	12.	T

Part E

IRONS

—NOTES—

Chapter 15

CAST IRON AND ITS ALLOYS

Cast iron is a mixture of 91 to 94 percent iron and varying proportions of other elements, the most important being carbon and the others as follows:

- Silicon
- Manganese
- Sulfur
- Phosphorous

The term cast iron refers to a class of metals sharing most of the same characteristics. The four basic types of cast iron are:

- Gray cast iron
- White cast iron
- Malleable cast iron
- Nodular cast iron

Cast-iron welding is useful in correcting casting defects, in repairing worn or broken castings, and in fabricating assemblies. Machine-support frames, large machine bases, housings, enclosure assemblies, and pipe fittings are typical applications for cast-iron welding.

CAST-IRON ALLOYS

Various improvements in cast iron can be made by adding nickel, molybdenum, chromium, copper, aluminum, and other elements to gray cast iron. This produces an alloy possessing characteristics that are greater in corrosion resistance, higher in strength, and so forth. For instance, a nickel-cast-iron alloy will have greater corrosion and wear resistance than gray cast iron. A molybdenum-cast-iron alloy will posses higher tensile strength.

Note: When welding alloy cast irons, the welder should take into consideration the characteristics of the alloying element and proceed accordingly. With aluminum cast irons, one problem is that aluminum oxides form on the surface during welding.

GRAY CAST IRON

Gray cast iron (also sometimes referred to as gray iron or pig iron) is an alloy composed of:

- Iron
- Carbon
- Silicon

Traces of phosphorous are also usually present. Chemical analysis of a gray cast iron will vary in accordance with its composition. For instance, the silicon content may vary from 1.0 to 3.0 percent, and the carbon content from 2.5 to 4.0 percent (in the free or graphitic state). The amounts will vary according to the intended use for the gray cast iron. Gray cast iron contains carbon of approximately 3.00 to 3.75 percent. Tensile strength of gray cast iron ranges from 10,000 to 60,000 psi, depending on the particular class.

Note: The American Society for Testing and Materials (ASTM) has devised a classification system based on the tensile strengths of different gray cast irons. A gray cast iron having a minimum tensile strength of 10,000 psi will belong to class 10; one with a minimum tensile strength of 20,000 psi, to class 20; and so forth. There are a total of seven classes.

The melting point of gray cast iron is approximately 2150 degrees F. Its ductility is rather low. Ductility is indicated by limited distortion at breaks. Impact and shock resistance are almost nonexistent. Gray cast irons are usually easy to machine and weld. They are also easily cast in a wide variety of forms.

Note: Gray cast iron is nonmalleable at any temperature.

MALLEABLE CAST IRON

Malleable cast iron is produced by annealing or heating white cast iron over a prolonged period of time and then allowing it to slowly cool down. This type of heat treatment produces a greater resistance to impact and shock, higher strength, and greater ductility. As a result, it is able to withstand various types of strain better than gray cast iron. Here are some characteristics:

- Tensile strength of malleable cast irons ranges from 30,000 to 100,000 psi.
- The strength depends upon whether the normal base structure is ferrite or pearlite.
- The lower tensile strengths (53,000 psi or below) belong to the ferrite malleable cast irons.
- The higher tensile strengths (60,000-100,000 psi) belong to the pearlite group.
- Most malleable cast iron is produced with a tensile strength of 40,000-50,000 psi.

It should be noted that as the tensile strength of a malleable cast iron increases, the ductility decreases.

Carbon content of malleable cast iron is approximately 2 to 3 percent and must be in the combined form. (Free or graphitic carbon cannot be malleableized.)

Malleable cast iron is stronger and tougher than gray cast iron. It possesses good machining characteristics, but because of malleableizing it has limited weldability.

Recommended joining processes for malleable cast iron are oxyacetylene welding and braze welding. Braze welding is preferred because of its lower temperatures. High welding temperatures trigger the malleable characteristic of being easy to break down. A reversion to the characteristics of white cast iron is the result. For this reason, welding processes that generate a high heat, such as shielded metal-arc, carbon-arc, and so forth, are not recommended for working with this metal unless temperatures of approximately 1400 to 1450 degrees F or below are used.

NODULAR CAST IRON

Nodular cast iron (also referred to as ductile cast iron or spheroidal cast iron) is so named because the graphite present in its composition takes the form of nodules rather than flakes. This results in greater shock resistance than with gray cast iron. The formation of graphite nodules is produced by adding magnesium to the molten metal during the production process.

The melting point of nodular cast iron is slightly below that of gray cast iron. Special heat treatment and the addition of alloying elements result in a metal that combines the more desirable characteristics of both gray cast iron and steel. The main alloying agent added to obtain these characteristics is magnesium.

The tensile strengths of nodular cast iron ranges from 60,000 to 120,000 psi, depending on the composition of the base metal. The process used to obtain the nodule structure and other factors play a role.

Note: Both oxyacetylene welding and oxyacetylene braze welding can be used to weld nodular cast iron.

WHITE CAST IRON

White cast iron is also referred to as high-strength cast iron. It is produced from pig iron when it cools very rapidly. The rate of cooling is too rapid to allow the carbon to separate from the iron-carbide compound, so the carbon found in white cast iron exists in the combined form.

White cast iron is very hard and brittle. It does not lend itself well to machining. As a result, this type of cast iron is not often used in its original state for castings. Due to its hardness, it is frequently used as a wear-resistant outer surface for inner cores of gray cast iron. These white cast-iron surfaces are produced by causing the molten metal to flow against heavy iron chills placed in the mold. This causes rapid cooling, which in turn results in a very hard surface. Castings produced in this manner are referred to as chilled-iron castings. One of the major uses of white cast iron is the production of malleable iron castings.

In most instances, white cast iron has a tensile strength of 40,000-50,000 psi or more. The melting point is 2300 degrees F, slightly higher than that of gray cast iron.

WELDING AND JOINING CAST IRON

Cast iron is less weldable than low-carbon steel, due to the fact that it contais greater amounts of carbon and silicon. As a result, it tends to be more brittle than low-carbon steel and has a tendency to crack if it is in the heat-affected zone unless proper preheating, postcooling, and low welding temperatures are used.

Note: The heat-affected zone (often abbreviated HAZ) is the area of the base metal along the joint seam that has not melted but whose structure has been altered by the welding or cutting heat.

Proper edge or area preparation is very important when welding cast iron. All surface defects must be completely removed. Sufficient space must be provided in the joint for satisfactory filler-metal deposition with the minimum penetration allowed. To minimize cracking, a ductile material should be chosen as a filler metal.

Common practice is to apply preheating to castings and to provide protected slow cooling after welding. Slow cooling is used to reduce residual stresses and avoid cracking. Preheating slows the cooling rate, permitting the formation of less brittle structures. It also permits the whole casting to contract with the weld material, thereby reducing residual stresses.

Welds in cast iron, if of sufficient thickness, may be strengthened by the mechanical method of studding. Studs of steel and approximately 1/4 to 3/8 inch in diameter should be used. The cast iron should be beveled to form a V shape. It should be drilled and tapped along the V so that the studs may be screwed into the casting. The studs should project about 3/16 to 1/4 inch above the cast-iron surface. They should be long enough to be screwed into the casting to a depth of at least the diameter of the studs.

Oxyacetylene Welding

The oxyacetylene-welding process can be used to weld both gray and nodular cast irons in all thicknesses. It is not recommended for welding malleable cast iron because it tends to promote brittleness.

Shielded-Metal-Arc Welding

Shielded metal-arc welding (SMAW) is commonly used to weld large gray cast-iron castings. Either a machinable or nonmachinable electrode can be used.

There are two basic types of machinable electrodes: a 100 percent nickel-core electrode and a nickel-iron base electrode. Both types produce a relatively soft, ductile weld deposit that can be machined after cooling.

These machinable electrodes are commonly used to repair broken or cracked castings, fill surface voids, or weld dissimilar metals, such as cast iron to steel

Nonmachinable electrodes, on the other hand, have a mild steel core. Because the electrode coating melts at low temperatures, a low welding current can be used to weld the cast iron. It produces a very hard and waterproof weld deposit that is not machined afterwards. This electrode is used for repairing automotive engine blocks, compressor blocks, water jackets, pump parts, and other components that require a tough joint without the need for machining.

Gas-Tungsten-Arc Welding

Gas-tungsten arc welding (GTAW), or TIG welding, can be used to weld gray cast iron, but it does not provide any specific advantages over less expensive welding processes. A higher preheat is usually recommended. Filler materials in the form of rods but of chemical composition similar to those used for SMAW can be used.

Gas-Metal-Arc Welding

Gas-metal-arc welding (GMAW), or MIG welding, is recommended for welding ductile and malleable cast iron, although it can be used for welding gray cast iron when productivity is important. GMAW produces high deposition rates with limited weld penetration. It is recommended for welding thicknesses of 1/4 inch or thicker.

Flux-Cored-Arc Welding

Flux-cored-arc welding (FCAW) is a wire-feed welding process in which a continuous consumable electrode wire is fed into the molten weld pool. FCAW produces high deposition rates with limited weld penetration. It is recommended for welding thicknesses of 1/4 inch or thicker.

Braze Welding

One very common use of braze welding is in the repair of gray iron castings. A copper-base alloy is generally used to either join fractured parts or to build up worn or missing sections.

Braze welding with copper-alloy filler metals is actually very effective on any type of cast iron. The low temperature requirements of braze welding make the process particularly suitable for joining malleable iron.

Among the problems associated with the use of the braze-welding process for welding cast iron are:

- The color of the copper-alloy filler metal does not match that of the iron.

- The corrosion resistance of the weld metal differs from that of the base metal.

- Galvanic corrosion may become a problem after dissimilar metals are joined.

- The strength of a braze weld falls off rapidly as temperatures increase.

Preheating cast iron may be local or it may extend over a wide area. It all depends on the size of the casting. Large castings require extensive preheating.

The preheating temperature is important. If the preheat temperature is too hot or too cold, the filler metal will not wet (tin) the joint. Cast iron seldom requires preheating to more than 1000 degrees F, and lower temperatures are often used. The base metal will determine the correct preheating temperature. Note: The heat added during braze welding is normally of short duration.

Brazing

Torch-brazing cast iron is used in limited production and/or repair and maintenance applications. It is used to join cast iron by heating the metal to a temperature above 800 degrees F. The filler metal then flows between the closely fitted adjoining surfaces by capillary attraction. The type of brazing used to join cast iron is called silver-alloy brazing because it uses an alloy of silver, copper, and/or zinc mixed with small amounts of other alloying elements.

Soldering

Torch soldering is used to join cast iron by heating the metal to a temperature below 800 degrees F. Most soldering of cast iron is done to repair castings or to fill surface imperfections. Although cast irons are generally difficult to solder, soldering has been used to join gray, malleable, and nodular cast iron ranging in thicknesses from 1/8 to 3/4 inch.

Note: Gray cast iron is particularly difficult to solder because its graphitic-iron content prevents the molten solder from bonding properly. White cast iron is almost never soldered.

MULTIPLE-CHOICE EXAM

1. Cast iron is a mixture of 91 to 94 percent iron and varying proportions of other elements. What is the most important element?

    a. silicon   **b.** manganese

    c. sulfur   **d.** carbon

2. The term cast iron refers to a class of metals sharing most of the same characteristics. Which of the following is not one of them?

    a. red cast iron   **b.** white cast iron

    c. malleable cast iron   **d.** nodular cast iron

3. Typical cast-iron applications are:

    a. machine support frames   **b.** large machine bases

    c. housings and pipe fittings **d.** all of the above

    e. none of the above

4. Various improvements in cast iron can be made by adding nickel, molybdenum, chromium, copper, aluminum, and other elements to gray cast iron. Which one is added to gain greater corrosion protection and wear resistance?

    a. chromium   **b.** aluminum

    c. copper   **d.** nickel

5. Gray cast iron is an alloy composed of iron, carbon, and _____.

    a. chromium   **b.** nickel

    c. copper   **d.** silicon

6. Malleable cast iron with high tensile strength belongs to the _____ group.

    a. pearlite   **b.** cast

    c. ferrite   **d.** steel

7. Gray cast irons are usually easy to machine and _____.

    a. file   **b.** cut

    c. weld   **d.** solder

8. The carbon content of malleable iron is between 2 and _____ percent.

    a. 10   **b.** 3

    c. 5   **d.** 20

9. Nodular cast iron is also sometimes called _____ iron.

    a. circular   **b.** spheroidal

    c. white   **d.** gray

    e. ductile cast iron

10. White cast iron is very hard and _____.

    a. brittle   **b.** weak

    c. strong   **d.** flaky

11. White cast iron is also referred to as high- _____ cast iron.

    a. strength   **b.** crust

    c. ductile   **d.** carbon

12. Cast iron is _____ weldable compared to low-carbon steel.

    a. highly   **b.** extremely

    c. more   **d.** less

13. The heat-affected zone is the area of the base metal along the joint seam that has not melted but whose structure has been altered by the welding or cutting _____.

    a. flame   **b.** tip

    c. heat   **d.** torch

14. Welds in cast iron, if of sufficient thickness, may be strengthened by the mechanical method of _____.

    a. bridging   **b.** studding

    c. fillets   **d.** riveting

15. Shielded-metal-arc welding (SMAW) is commonly used to weld large gray-iron _____.

    a. castings   **b.** pipe

    c. fixtures   **d.** butt joints

16. The nonmachinable electrode has a _____ -steel core.

    a. hard   **b.** soft

    c. mild   **d.** none of the above

17. TIG welding can be used to weld _____ cast iron.

    a. mild   **b.** nodular

    c. white   **d.** gray

18. MIG welding is recommended for welding ductile and _____ cast iron.

    a. white   **b.** malleable

    c. gray   **d.** nodular

19. One very common use of braze welding is in the repair of _____ iron castings.

    a. nodular   **b.** white

    c. gray   **d.** malleable

20. Gray cast iron is particularly difficult to _____ because its graphitic-iron content prevents the molten filler metal from bonding properly.

    a. braze   **b.** solder

    c. flux   **d.** weld

TRUE-FALSE EXAM

1. Cast-iron welding is useful in correcting casting defects and repairing worn or broken castings.

True   False

2. When welding alloy cast irons, the welder should take into consideration the characteristics of the alloying element and proceed accordingly.

True   False

3. Gray iron is also called pig iron.

True   False

4. Chemical analysis of a gray cast iron will vary in accordance with its composition.

True   False

5. Gray cast iron is malleable at any temperature.

True   False

6. Malleable cast iron is produced by annealing or heating white cast iron over a prolonged period and then allowing it to slowly cool down.

True   False

7. The melting point of gray cast iron is approximately 2400 degrees F.

True   False

8. Gray cast irons are usually easy to machine and weld.

True   False

9. Most malleable iron is produced with a tensile strength of 40,000 to 50,000 psi.

True   False

10. It should be noted that as the tensile strength of a malleable cast iron increases, the ductility also increases.

True   False

11. Malleable cast iron is stronger and tougher than gray cast iron.

True   False

12.   Recommended joining processes for working with malleable cast iron are oxyacetylene welding and braze welding.

True   False

13.   Nodular cast iron is also referred to as ductile cast iron or spheroidal cast iron.

True   False

14.   Both oxyacetylene welding and oxyacetylene braze welding can be used to weld nodular cast iron.

True   False

15.   White cast iron is very hard and brittle.

True   False

16.   The melting point of white cast iron is 2300 degrees F.

True   False

17.   Cast iron is less weldable than low-carbon steel.

True   False

18.   The melting point of white cast iron is slightly lower than that of gray cast iron.

True   False

19.   Oxyacetylene welding can be used to weld both gray and nodular cast irons in all thicknesses.

True   False

20.   One very common use of braze welding is in the repair of gray iron castings.

True   False

MULTIPLE-CHOICE ANSWERS

1.	D	5.	D	9.	B	13.	C	17.	D
2.	A	6.	A	10.	A	14.	B	18.	B
3.	D	7.	C	11.	A	15.	A	19.	C
4.	D	8.	B	12.	D	16.	C	20.	B

TRUE-FALSE ANSWERS

1.	T	6.	T	11.	T	16.	T
2.	T	7.	F	12.	T	17.	T
3.	T	8.	T	13.	T	18.	F
4.	F	9.	T	14.	T	19.	T
5.	F	10.	F	15.	T	20.	T

—NOTES—

Chapter 16
WROUGHT IRON

Wrought iron is iron with little or no carbon. It has a spark stream that is composed of small particles that flow away from the wheel in a straight line. The sparks become wider and brighter as they get further away from the wheel until finally they go out. Wrought iron was the type of iron used before steel could be produced economically.

Wrought iron is produced in a puddling furnace by melting pig-iron ingots. This melting process removes almost all of the carbon and other impurities such as silicon and manganese. The highly refined iron combines in a mechanical mixture with the iron-oxide-silicate slag to produce the fibrous structure of wrought iron. Keep in mind that wrought iron has:

• Good tensile strength

• High ductility

• High corrosion resistance

• Fatigue resistance

Wrought iron has a tensile strength of about 45,000 psi. Its melting point is approximately 2800 degrees F, or several hundred degrees above the melting point of its own slag.

Wrought iron is very easy to work with because it approaches its melting point gradually and is easy to control. Wrought iron can be welded quite satisfactorily with all-purpose steel welding rods. There are also low-carbon-content-steel welding rods that have been designed specifically for welding wrought iron. This form of iron is characterized by its extremely low carbon content. In fact, it has the lowest carbon content of all the various commercial irons.

Welding and Joining Processes

Either oxyacetylene or braze welding can be used to weld wrought iron. Braze-welded joints are not nearly as strong as those produced by oxyacetylene welding.

Oxyacetylene Welding

Oxyacetylene welding has long been the preferred welding process used to bond wrought iron. However, these welds do not have a wrought-iron structure. Here are some tips for working with oxyacetylene:

• Use a neutral oxyacetylene flame to weld wrought iron.

• Use either a low-alloy-steel gas-welding rod (RG 60) or a low-carbon-steel gas welding rod (RG 45).

• Do not use high-carbon-steel welding rods to increase weld strength. The results are not very satisfactory.

• Avoid excessive stirring of the molten-weld puddle. Doing so often results in oxide formation, caused by oxygen combining with elements in the weld puddle.

No flux is required when welding wrought iron. Excellent protection against oxidation is provided for the weld and adjacent surfaces because the slag melts at a temperature considerably below that of the wrought-iron base metal.

Joint design includes the following elements:

- Lap joint

- T-joint

- V-groove butt joint

- Fillet joints

- Plug joint

Note: Slag sweats to the surface at a relatively low temperature. And, when welding wrought iron with the oxyacetylene process, it is a false indicator of fusion. Keep the welding rod immersed in the molten puddle. Continue to apply heat until the base metal breaks down and fusion is complete.

Braze Welding

In braze welding, the filler metal melts above 842 degrees F. This is below the melting point of the base metal. In other words, no fusion of the base metal takes place as it does in oxyacetylene welding. Unlike brazing or soldering, the filler metal is not distributed in the joint by capillary action. It is deposited in the joint, where it adheres to the base metal. Although the joint formed is not as strong as one formed by oxyacetylene welding, the low temperatures used in braze welding produce less stress and distortion. Note the following in the wrought-iron braze-welding process:

- Use a neutral oxyacetylene flame to braze-weld wrought iron.

- Use a bronze gas-welding rod (RB CuZn-A), a nickel gas-welding rod (RB CuZn-B), or a manganese bronze gas-welding rod (R Cu Zn-C).

- Use a flux (either a liquid paste or a powder) designed for braze-welding cast iron or wrought iron.

- Joint design is similar to that used in oxyacetylene welding.

MULTIPLE-CHOICE EXAM

1.   Wrought iron is an iron with little or no _____.

   a. carbon   **b.** manganese

   c. copper   **d.** slag

2.   Wrought iron is produced in a _____ furnace.

   a. reduction   **b.** puddling

   c. distilling   **d.** cast iron

3.   Wrought iron has good tensile _____.

   a. strength   **b.** ability

   c. weldability   **d.** none of the above

4.   The melting point of cast iron is about _____ degrees Fahrenheit.

   a. 1750   **b.** 1800

   c. 2800   **d.** 3200

5.   Two types of gas welding are used with wrought iron: oxyacetylene and _____ welding.

   a. braze   **b.** arc

   c. cast iron   **d.** none of the above

6.   No flux is required when welding _____ iron.

   a. cast   **b.** white

   c. gray   **d.** wrought

7.   What is a false indicator of fusion in the welding of wrought iron?

   a. flux   **b.** silicon

   c. slag   **d.** none of the above

8.   Which type of flame is used in welding wrought iron?

   a. neutral   **b.** blue

   c. white   **d.** none of the above

9. Joint design for wrought-iron welding is similar to that used in _____ welding.

    a. spot   **b.** oxyacetylene

    c. arc   **d.** none of the above

10. In braze welding the filler metal melts above _____ degrees Fahrenheit.

    a. 842   **b.** 950

    c. 1082   **d.** none of the above

TRUE-FALSE EXAM

1. Wrought iron is an iron with little or no carbon.

    True   False

2. Wrought iron was the type of iron used before steel became popular.

    True   False

3. Wrought iron is produced in a pig-iron furnace.

    True   False

4. Keep in mind that wrought iron does not have good tensile strength.

    True   False

5. The melting point of wrought iron is about 2800 degrees Fahrenheit.

    True   False

6. Either oxyacetylene or braze welding can be used to weld wrought iron.

    True   False

7. Wrought iron can be welded quite satisfactorily with all-purpose-steel welding rods.

    True   False

8. No flux is required when welding wrought iron.

    True   False

9.   When welding wrought iron with the oxyacetylene process, slag sweating to the surface gives a false indication of fusion.

   True   False

10. Do not use a neutral flame to braze-weld wrought iron.

   True   False

MULTIPLE-CHOICE ANSWERS

1. A	3. A	5. A	7. C	9. B
2. B	4. C	6. D	8. A	10. A

TRUE-FALSE ANSWERS

1. T	4. F	7. T	9. T
2. T	5. T	8. T	10. F
3. F	6. T		

Part F

STEELS

—NOTES—

Chapter 17
CARBON STEELS

Carbon and low-carbon steels are the workhorse materials for:

- Construction equipment
- Transportation equipment
- Industrial equipment
- Consumer products of many types

They comprise over 90 percent of total steel production. More carbon steel is used in product manufacture than all other metals combined.

It is not always easy to make a clear-cut distinction between iron and steel. This is particularly true if the carbon content of either one is close to the 2 percent level, because this is frequently used as a factor in classifying the two.

If the carbon content exceeds 2 percent, the metal is classified as one of the several iron types: cast iron, wrought iron, and so forth. On the other hand, a carbon content of less than 2 percent indicates that the metal should be classified as steel. For some common uses of carbon steel see Table 17.1.

TABLE 17.1 Carbon steel applications.

Carbon class	Carbon range %	Typical uses
Low	0.05–0.15	Chain, nails, pipe, rivets, screws, sheets for pressing and stamping, wire.
	0.15–0.30	Bars, plates, structural shapes.
Medium	0.30–0.45	Axles, connecting rods, shafting.
High	0.45–0.60	Crankshafts, scraper blades.
	0.60–0.75	Automobile springs, anvils, bandsaws, drop hammer dies.
Very high	0.75–0.90	Chisels, punches, sand tools.
	0.90–1.00	Knives, shear blades, springs.
	1.00–1.10	Milling cutters, dies, taps.
	1.10–1.20	Lathe tools, woodworking tools.
	1.20–1.30	Files, reamers.
	1.30–1.40	Dies for wire drawing.
	1.40–1.50	Metal cutting saws.

LOW-CARBON STEELS

Low-carbon steels (also referred to as mild steels or mild carbon steels) are characterized by a carbon content not exceeding 0.30 percent. The minimum amount of carbon found in these steels ranges from 0.05 to 0.08 percent. A carbon content below these minimum levels (i.e., below 0.05 percent) indicates a type of pure iron referred to as ingot iron.

Note: Low-carbon steels, with less than 0.25 percent carbon and a sulfur and phosphorus content less than 0.04 percent, can be easily welded with most of the welding processes. No special precautions need to be taken. The welding heat will not affect their basic properties.

Low-carbon steels have a melting point of approximately 2600 to 2700 degrees F. The slag contained in these steels will melt at a lower temperature than the metal itself, making the use of a flux unnecessary. Low-carbon steels possess a number of excellent qualities that contribute to their being the most widely used steels in industry today. First and probably foremost, they are relatively inexpensive. They are also tough and ductile. Finally, they are generally easy to weld and machine.

Note: Copper-bearing steels are low-carbon steels that contain approximately 0.20 percent copper. A copper content exceeding this amount generally causes surface cracking in the base metal surrounding the weld. Copper-bearing steels are welded with the same procedures used for other low-carbon steels.

Following are summary descriptions of welding types used with low-cabon steels:

- Shielded-Metal-Arc Welding

    a. Shielded-metal-arc welding is commonly used to weld mild steel in low-carbon ranges. As the carbon content increases, it becomes necessary to apply preheat and postheat treatment procedures in order to obtain a satisfactory weld.

    b. Both mild-steel electrodes (E60XX) and low-hydrogen electrodes (e.g., E6015 or E6016) are recommended for use with low-carbon steels.

    c. Weld cracking can be avoided when working with medium-carbon steels in the higher carbon ranges by employing special welding techniques and electrodes suitable for this purpose.

    d. A postheat treatment is recommended.

- Oxyacetylene Welding

    a. Oxycetylene welding is commonly used to weld thin gauges of sheets or tubes.

    b. Due to the significant difference between the melting temperature of the mild steel and its slag, a flux is generally not used. The selection of a filler metal, however, is important and will depend on the analysis of the base metal. This in turn will determine the type of flame used.

    c. There are three ferrous welding rods recommended as filler metals for the low-carbon (mild) steels. These are:
       (1) An alloy-steel rod (AWS Classification GA 60)
       (2) A high-tensile-steel rod (AWS Classification GA 60)
       (3) A low-carbon (mild)-steel rod (AWS Classification GA 50)

    d. The alloy-steel rod will produce a weld with a tensile strength in excess of 60,000 psi. The weld produced by the low-carbon-steel rod will be somewhat lower, with a minimum tensile strength of 52,000 psi.

    e. A neutral flame or one with a slight excess of acetylene is recommended when using the alloy-steel rod. A slightly carburizing flame is also recommended for welding with either a high-tensile- or low-carbon-steel rod (with the carburizing flame a somewhat higher welding speed will be achieved). An oxidizing flame must be strictly avoided, because it will form oxides on the surface that may be included in the weld and thereby weaken it. Excessive sparking is one indicator of an oxidizing flame.

f. A slightly carburizing flame is also recommended for welding with either a high-tensile- or low-carbon-steel rod. A somewhat higher welding speed will result by using the carburizing flame.

g. An oxidizing flame must be strictly avoided, because it will form oxides on the surface that may be included in the weld and thereby weaken it. Excessive sparking is one indicator of an oxidizing flame.

h. A backhand welding technique is preferred to a forehand one. In backhand welding the flame is directed back against the weld, where it functions as a shield against contamination and thus reduces the possibility of oxide formation. Using this technique also results in less torch and rod manipulation, less agitation of the metal, and faster welding speeds.

i. Because of the significant difference between the melting temperature of the mild steel and its slag, a flux is generally not used.

j. The selection of a filler metal is important and will depend on the analysis of the base metal. This, in turn, will determine the type of flame used.

k. A postheat treatment is recommended

MEDIUM-CARBON STEELS

The carbon content of medium-carbon steels ranges from 0.30 to 0.45 percent. These steels are not as easily welded as low-carbon steels, which is directly due to the increased carbon content. Medium-carbon steels are harder and stronger than mild steels but tend toward weld brittleness in the high-carbon range.

Keep in mind the following:

- Low-hydrogen electrodes are recommended when using the SMAW process to weld medium-carbon steels.

- The E-6015, E-6016, and E-6018 electrodes require preheating the base metal to reduce weld cracking. These electrodes are used where greater amounts of weld deposit are required.

HIGH-CARBON STEELS

High-carbon steels contain 0.45 to 0.75 percent carbons. They find common application in the manufacture of such items as knives, saws, bits, and tools. In other words, these manufactured items require hard, tough surfaces exhibiting a high resistance to wear. Some classifications of the carbon steels add a fourth subgroup with an even higher carbon-content range (0.75 to 1.50 percent). These are referred to as very-high-carbon steels. Due to their extreme hardness and other characteristics that limit the weld quality, these carbon steels are seldom welded except for repair purposes.

Note the following:

- The major problem with high-carbon steels (and very-high-carbon steels) is that they require controlled preheating and postheating. The preheating and postheating phases must be maintained within a specific temperature range. The following temperature ranges are recommended:

   a. Preheat steel with a carbon content of 0.45 to 0.65 percent from 200 to 400 percent F (94 to 204 percent C)

   b. Preheat steel with a carbon content of more than 0.65 percent from 752 to 1292 percent F (204 to 371 percent C).

   c. Postheat all high-carbon steels from 1098 to 1200 degrees F (592 to 650 degrees C).

- The adverse effects of the welding heat can be somewhat lessened by maintaining a fast welding speed.

- Low-hydrogen electrodes are recommended when welding high-carbon steels with the SMAW process. These electrodes will greatly reduce the possibility of cracking.

- Many high-carbon steels require post-heat treatment for stress relief.

TABLE 17.2   Prepared analyses for steels to be arc welded.

Element	Composition preferred	High*
Carbon	0.06–0.25	0.35
Manganese	0.35–0.80	1.4
Silicon	0.10 or less	0.30
Sulfur	0.035 or less	0.05
Phosphorus	0.030 or less	0.04

* Additional care is required in welding of steels containing the amounts of the elements listed.
   (Lincoln Electric)

MULTIPLE-CHOICE EXAM

1. Which of the following is not one of the work-horse materials of carbon steels?

    a. construction equipment   **b.** transportation equipment

    c. industrial equipment   **d.** aviation equipment

2. A carbon content of less than 2 percent indicates that the metal should be classified as _____.

    a. steel   **b.** cast iron

    c. gray iron   **d.** nodular iron

3. Low-carbon steels are also known as _____ steels.

    a. mild   **b.** ingot iron

    c. gray iron   **d.** white iron

4. Low-carbon steels with less than 0.25 percent carbon and a sulfur and phosphorus content less than _____ percent can be easily welded with most of the welding processes.

    a. 0.02   **b.** 0.03

    c. 0.04   **d.** 0.05

5. Low-carbon steels have a melting point of _____- degrees Fahrenheit.

    a. 1200 to 1400   **b.** 1400 to 1500

    c. 2500 to 2600   **d.** 2600 to 2700

6. Low-carbon steels are generally easy to weld and _____.

    a. melt   **b.** solder

    c. braze   **d.** machine

7. As the carbon content increases, it becomes necessary to apply preheat and _____ treatment procedures in order to obtain a satisfactory weld.

    a. low-heat   **b.** postheat

    c. high-heat   **d.** none of the above

8. Oxyacetylene welding is used to weld thin gauges of sheets or _____.

    a. tubes   **b.** stainless steel

    c. tanks   **d.** none of the above

9. An oxidizing flame must be strictly avoided because it will form _____ on the surface that may be included in the weld and thereby weaken it.

    a. nitrogen   **b.** hydrogen

    c. oxides   **d.** none of the above

10. High-carbon steels contain 0.45 to _____ percent carbons.

    a. 0.04   **b.** 0.25

    c. 0.50   **d.** 0.75

TRUE-FALSE EXAM

1. Carbon and low-carbon steels are the workhorse materials for consumer products of many kinds.

True   False

2. More carbon steel is used in product manufacture than all other metals combined.

True   False

3. If carbon content exceeds 5 percent, the metal is classified as one of the several irons-cast iron, wrought iron, and so forth.

True   False

4. Shielded metal-arc is commonly used to weld mild steel in the low carbon ranges.

True   False

5. Very high carbon content in steel makes it ideal for making dies for wire drawing and files.

True   False

6. The slag content in low-carbon content will melt at a lower temperature than the metal itself.

True   False

7.   Flux is unnecessary in the welding of low-carbon steels.

     True   False

8.   Copper-bearing steels are low-carbon steels that contain approximately 0.20 percent copper.

     True   False

9.   Oxyacetylene welding is commonly used to weld thick-gauged sheets of low-carbon steel.

     True   False

10.   A slightly carburizing flame is not recommended for welding with either a high-tensile or low-carbon steel rod.

     True   False

MULTIPLE-CHOICE ANSWERS

1.	D	3.	A	5.	D	7.	B	9.	C
2.	A	4.	C	6.	D	8.	A	10.	D

TRUE-FALSE ANSWERS

1.	T	4.	T	7.	T	9.	T
2.	T	5.	T	8.	T	10.	F
3.	F	6.	T				

—NOTES—

Chapter 18

LOW- AND HIGH-ALLOY STEELS

Steel, in one form or another, is the most used metal for the do-it-yourself welder and for structural and construction purposes. Therefore, the welder must know the maerials that he or she is working with before acquiring the proper equipment to work these metals for whatever purpose they are to be put. One place to start is by acquiring some knowledge of the steels to be welded. Some of the facts to keep in mind are discussed in the following paragraphs.

Alloy steel is categorized as either high-strength or low-alloy. Both rely on chemical composition to develop the desired mechanical properties. The construction alloy steels develop their desired properties by thermal treatment. Many manufactured products require steels with good hardening ability, ductility, and fatigue strength-mainly the constructional alloys. For structural application, however, high yield strength, good weldability, and corrosion resistance are most desired, with only limited ductility and virtually no hardening ability needed. These steels have resulted in substantial weight savings in automobiles, trains, bridges, and buildings.

Low-alloy structural steels have about twice the yield strength of plain-carbon structural steels. This increase in strength, coupled with resistance to martensite formation in a weld zone, is obtained by adding low percentages of several elements, notably manganese, silicon, niobium (columbium), and vanadium, as well as several others.

Alloys of steel start with the steel itself. Steel is an alloy of iron and carbon. When produced, certain other elements are present in small quantities:

- Manganese
- Phosphorous
- Sulfur
- Silicon

When these four foreign elements are present in their normal percentages, the product is referred as plain-carbon steel. Its strength is primarily a function of its carbon content. The ductility of plain-carbon steel decreases as carbon content is increased. Its hardening ability is quite low. In addition, the properties of ordinary carbon steels are impaired by both high and low temperatures. They are subject to corrosion in most environments.

DIFFERENCES IN STEELS

The differences between plain-carbon and alloy steel are often debatable. Both contain carbon, manganese, and usually silicon. Copper and boron also are possible additions to both classes. Steels containing more than 1.65 percent manganese, 0.60 percent silicon, or 0.60 percent copper are designated as alloy steels. Steel is also considered to be an alloy steel if a definite amount or a minimum of another alloying element is specified or required. The most common alloy elements are chromium, nickel, molybdenum, vanadium, tungsten, cobalt, boron, and copper, as well as manganese, silicon, phosphorous, and sulfur in amounts greater than normally are present. (See Table 18.1.)

TABLE 18.1 Alloying elements for steel.

Element	Percentage	Primary Function
Manganese	0.25–0.40 >1%	Combine with sulfur to prevent brittleness Increase hardenability, by lowering transformation points and causing transformations to be sluggish
Sulfur	0.08–0.15	Free-machining properties
Nickel	2–5 12–20	Toughener Corrosion resistance
Chromium	0.5–2 4–18	Increase hardenability Corrosion resistance
Molybdenum	0.2–5	Stable carbides; inhibits grain growth
Vanadium	0.15	Stable carbides; increases strength while retaining ductility; promotes fine grain size
Boron	0.001–0.003	Powerful hardenability agent
Tungsten		Hardness at high temperatures
Silicon	0.2–0.7 2 Higher percentages	Increases strength Spring steels Improve magnetic properties
Copper	0.1–0.4	Corrosion resistance
Aluminum	0.95–1.30	Alloying element in nitriding steels
Titanium	—	Fixes carbon in inert particles Reduces martensitic hardness in chromium steels
Lead	—	Improves machinability

PLAIN-CARBON STEEL

Plain-carbon steels are classified into subgroups:

- Low-carbon steels
- Medium-carbon steels
- High-carbon steels

Low-carbon steels have less than 0.30 percent carbon. Medium-carbon steels have between 0.30 and 0.80 percent carbon. High-carbon steels have more than 0.80 percent carbon. (See Chapter 17 for additional information.)

EFFECTS OF ALLOYING STEELS

Alloying elements are added to steel in small percentages—usually less than 5 percent. Alloying is used to improve strength or hardening ability. In much larger amounts, often up to 20 percent, the alloy produced has special properties such as corrosion resistance or stability at high or low temperatures. Additions may be made during the steel-making process to remove dissolved oxygen from the melt. Note the melting point in the various metals and alloys in Table 18.2. Here are some used for alloys:

- Manganese, silicon, and aluminum are often used for deoxidation.

- Aluminum and to a lesser extent vanadium, columbium, and titanium are used to control austenitic grain size.

- Sulfur, lead, selenium, and tellurium are used for machinability.

- Other elements may be added to improve the strength or toughness properties of the metal.

- Manganese, silicon, nickel, and copper add strength by forming solid solutions in ferrite.

- Chromium, vanadium, molybdenum, tungsten, and other elements increase strength by forming dispersed second-phase carbides.

- Columbium, vanadium, and zirconium can be used for ferrite grain-size control.

- Nickel and copper are added to low-alloy steels to provide improved corrosion resistance.

TABLE 18.2 Melting points of metals and alloys.

Carbon	3500	
	3400	
	3300	
	3200	
Chromium	3100	
Pure iron	3000	Wrought iron
Mild steel	2900	Stainless steel, 12%
	2800	chromium
	2700	Cobalt
Nickel	2600	Silicon
Stainless steel,	2500	
19% chromium	2400	
Manganese	2300	
	2200	Cast iron
	2100	
	2000	Copper
	1900	
Silver	1800	Brass
	1700	
	1600	Bronze
	1500	
	1400	
Aluminum	1300	Magnesium
	1200	
	1100	
	1000	Aluminum alloys
	900	Magnesium alloys
Zinc	800	
	700	
Lead	600	
	500	
Tin	400	

TABLE 18.3   Metal identification by appearance.

	Alloy Steel	*Copper*	*Brass and Bronze*	*Aluminum and Alloys*	*Monel Metal*	*Nickel*	*Lead*
Fracture	Medium gray	Red color	Red to yellow	White	Light gray	Almost white	White; crystalline
Unfinished surface	Dark gray; relatively rough; rolling or forging lines may be noticeable	Various degrees of reddish brown to green due to oxides; smooth	Various shades of green, brown, or yellow due to oxides; smooth	Evidences of mold or rolls; very light gray	Smooth; dark gray	Smooth; dark gray	Smooth; velvety; white to gray
Newly machined	Very smooth; bright gray	Bright copper red color dulls with time.	Red through to whitish yellow; very smooth	Smooth; very white.	Very smooth; light gray	Very smooth; white	Very smooth; white

	White Cast Iron	*Gray Cast Iron*	*Malleable Iron*	*Wrought Iron*	*Low-Carbon Steel and Cast Steel*	*High-Carbon Steel*
Fracture	Very fine silvery white silky crystalline formation;	Dark gray	Dark gray	Bright gray	Bright gray	Very light gray
Unfinished surface	Evidence of sand mold; dull gray	Evidence of sand mold; very dull gray	Evidence of sand mold; dull gray	Light gray; smooth	Dark gray; forging marks may be noticeable; cast-evidences of mold	Dark gray; rolling or forging lines may be noticeable
Newly machined	Rarely machined	Fairly smooth; light gray	Smooth surface; light gray	Very smooth surface; light gray	Very smooth; bright gray	Very smooth; bright gray

IDENTIFICATION OF STEELS

Welders should be able to identify the metals they are welding. Besides testing the metal with a magnet to determine if it is ferromagnetic, the physical color of the metal is a basis for determining its composition. Checking a metal's melting point gives a more accurate method of determining its composition. Most metals' melting points are available in a number of locations.

Identification Methods

Steels can be identified by a number of methods. Once the metal has been used and is about to be reused or welded for another purpose, it may be necessary to first investigate the ability of the steel to do the job. Three of the ways to identify metals are shown in Tables 18.3 and 18.4 as well as Figure 18.1. In the spark test, the sparks from a grinding wheel have distinct shapes and colors. After a bit of practice the welder is able to tell which steel is being tested.

High-Alloy Cast Irons

For high-alloy cast irons that are not to be heat-treated, the alloy elements are often selected to alter the properties by affecting the formation of graphite or cementite. This means modifying the morphology of the carbon-rich phase or simply strengthening the matrix material. Alloys are often added in small amounts to improve strength properties or wear resistance. High-alloy cast irons are often designed to provide corrosion resistance especially at high temperatures, as in the chemical industry.

Among the high-alloy cast irons are austenitic gray cast irons. They are quite common. They contain about 14 percent nickel, 5 percent copper, and 2.5 percent chromium. They possess good corrosion resistance to acids and alkalis at temperatures up to about 800 degrees C or 1400 degrees F.

Tool Steels

Tool steels are designed to provide wear resistance and toughness combined with high strength. They are basically high-carbon alloys. Chemistry provides the balance of toughness and wear desired for this type of steel.

Several classification systems have been applied to tool steels. (See Table 18.5.)

TABLE 18.4   Metal identification by chips.

	Copper	Brass and Bronze	Aluminum and Alloys	Monel Metal	Nickel	Lead
Appearance of chip	Smooth chips; saw edges where cut	Smooth chips; saw edges where cut	Smooth chips; saw edges where cut	Smooth edges	Smooth edges	Any shaped chip can be secured because of softness
Size of chip	Can be continuous if desired	Can be continuous if desired	Can be continuous if desired	Can be continuous if desired	Can be continuous if desired	Can be continuous if desired
Facility of chipping	Very easily cut	Easily cut; more brittle than copper	Very easily cut	Chips easily	Chips easily	Chips so easily it can be cut with penknife

	White Cast Iron	Gray Cast Iron	Malleable Iron	Wrought Iron	Low-Carbon Steel and Cast Steel	High-Carbon Steel
Appearance of chip	Small broken fragments	Small, partially broken chips but possible to chip a fairly smooth groove	Chips do not break short as in cast iron	Smooth edges where cut	Smooth edges where cut	Fine grain fracture; edges lighter in color than low-carbon steel
Size of chip		$1/8$ in.	$1/4-3/8$ in.	Can be continuous if desired	Can be continuous if desired	Can be continuous if desired
Facility of chipping	Brittleness prevents chipping a path with smooth sides	Not easy to chip because chips break off from base metal	Very tough, therefore harder to chip than cast iron	Soft and easily cut or chipped	Easily cut or chipped	Metal is usually very hard, but can be chipped

TABLE 18.5   Basic types of tool steel with SAE grades.

Type	AISI–SAE Grade	
1. Water hardening	W	
2. Cold work	O	Oil-hardening
	A	Air-hardening medium-alloy
	D	High carbon/high chromium
3. Shock resisting	S	
4. High speed	T	Tungsten base
	M	Molybdenum base
5. Hot work	H	H1–H19: chromium base
		H20–H39: tungsten base
		H40–H59: molybdenum base
6. Plastic mold	P	
7. Special purpose	L	Low alloy
	F	Carbon–tungsten

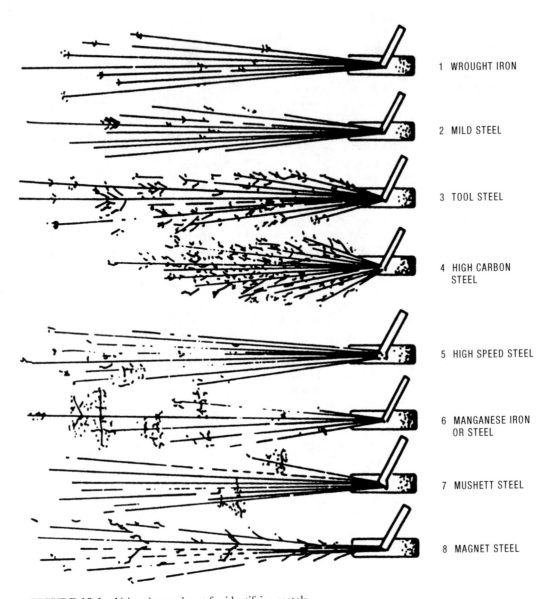

FIGURE 18.1 Using the spark test for identifying metals.

MULTIPLE-CHOICE EXAM

1. Desirable properties of steel are obtained by chemical _____.

    a. composition   **b.** addition

    c. purging   **d.** none of the above

2. Low-alloy structural steels have about _____ the yield strength of plain-carbon structural steels.

    a. twice   **b.** four times

    c. triple   **d.** none of the above

3. When manganese, phosphorous, sulfur, and silicon are present in their normal percentages, the product is referred to as _____ steel.

    a. plain-carbon   **b.** carbon

    c. alloy   **d.** none of the above

4. The most common alloy elements are chromium, nickel, molybdenum, vanadium, tungsten, cobalt, boron, and _____.

    a. carbon   **b.** silver

    c. mercury   **d.** copper

5. Low-carbon steels have less than _____ percent carbon.

    a. 0.50   **b.** 0.60

    c. 0.30   **d.** 0.40

6. Manganese, silicon, and aluminum are often used for _____ in alloying steel.

    a. deoxidation   **b.** strength

    c. weldability   **d.** hardening

7. Sulfur, lead, selenium, and tellurium are added for _____.

    a. softening   **b.** machinability

    c. hardening   **d.** appearance

8. The spark test produces distinct shapes and _____,

    a. qualities   **b.** colors

    c. sizes   **d.** none of the above

9. Among the high-alloy cast irons are austenitic _____ cast irons.

    a. white   **b.** red

    c. gray   **d.** nodular

10. Tool steels are designed to provide wear resistance and toughness combined with _____ strength.

    a. more   **b.** less

    c. low   **d.** high

TRUE-FALSE EXAM

1. Alloy steel is categorized as high-strength and low-alloy.

   True   False

2. Low-alloy structural steels have about twice the yield strength of plain-carbon structural steels.

   True   False

3. Steel is an alloy of iron and carbon.

   True   False

4. The ductility of plain-carbon steel increases as carbon content is increased.

   True   False

5. The difference between plain-carbon and alloy steel is often debatable.

   True   False

6. Steels containing less than 1.65 percent manganese, 0.60 percent silicon, or 0.60 percent copper are designated as alloy steels.

   True   False

7. Medium-carbon steels have between 0.30 and 0.80 percent carbon.

   True   False

8. Steels can be identified by a number of methods.

   True   False

9. In the spark test, the sparks from the grinding wheel have distinct shapes and colors.

   True   False

10. Among the high-alloy cast irons is austenitic gray cast iron.

    True   False

MULTIPLE-CHOICE ANSWERS

1.	A	3.	A	5.	C	7.	B	9.	C
2.	A	4.	D	6.	A	8.	B	10.	D

TRUE-FALSE ANSWERS

1.	T	4.	F	7.	T	9.	T
2.	T	5.	T	8.	T	10.	T
3.	T	6.	F				

Chapter 19

HARDFACING, TOOL, AND DIE STEELS

The more the welder knows about steel, the more valuable he/she is in the welding field. Working with metals and in some cases plastics, the welder must be versatile and eager to learn as the field progresses. With the advent of automatic and semiautomatic welding came the pressure on the welder to become more skilled and educated. By examining the material here the world of metals and metal processes becomes more familiar and useful.

CARBON STEEL

Carbon steel that has had other types of metal added to it to improve its properties is called an alloy steel. Increased strength and hardness are the usual properties desired. Retention of the metal's properties at elevated temperatures and improvement of the metal's resistance to corrosion are also desirable qualities. Elements added to steel for alloying are:

- Chromium
- Nickel
- Manganese
- Vanadium
- Molybdenum
- Silicon
- Tungsten
- Phosphorus
- Copper
- Titanium
- Zirconium
- Cobalt
- Columbium
- Aluminum

These elements form two groups. The first group forms a solid solution in which the element dissolves in iron. Silicon, manganese, nickel, molybdenum, vanadium, tungsten, and chromium increase the hardness and strength by strengthening the iron base of the alloy. The second group forms complex carbides when cooled slowly. These carbides precipitate out of a solid solution and are scattered throughout the metal. Chromium, manganese, vanadium, tungsten, titanium, columbium, and molybdenum are carbides of deep hardening. They increase wear resistance and the ability to withstand heat. Carbides are possible only when the base metal has sufficient carbon in the metal to form alloy carbides.

High-carbon steel has a carbon content ranging from 0.60 to 1.50 percent. It is very strong and hard. Such articles s as railroad equipment, automobile and truck parts, and farm machinery are made from high-carbon steel. Parts such as springs, grinding balls, bars, hammers, cables, axes, and wrenches are also produced from high-carbon steel with a content of 0.60 to 0.95 percent carbon. This grade of steel has great hardness, toughness, and strength.

When the carbon content is increased from 0.60 to 1.50 percent, the strength and wear resistance increase, but the hardness does not continue to increase. These steels are referred to as tool steels and are used to manufacture cutting and forming tools such as dies, taps, drills, reamers, chisels, and forming and bending dies. The maximum hardness is obtained with 0.80 percent carbon in steel. At 0.60 percent carbon, steel is near its maximum hardness.

HARDENING

Hardening metal is accomplished by increasing its temperature to its point of decalescence. Then it is quenched in a suitable cooling medium. Decalescence means a sudden increase in the rate of temperature rise of heated metal after a certain temperature has bee reached (1,463 degrees F, 795 degrees C for iron) due to a greater absorption of heat. In actual practice, the temperature of the metal that is to be hardened should be increased to slightly above the point of decalescence for two reasons:

- To be certain that the temperature of the metal is above the point of decalescence at all times

- To allow for a slight loss of heat while transferring the metal from the furnace to the quenching bath

Steel with less than 0.85 percent carbon needs to be heated above its upper critical temperature.

Heating Process

Steel can be heated most successfully by placing it in a cold furnace and then bringing the furnace and its charge to the hardening temperature slowly and uniformly. Commercially, only those steels that are difficult to harden are heat-treated in this manner. For example, some tool steels are heated for hardening by first placing them in a preheating furnace at a temperature of 1000 degrees F (537.77 degrees C) or slightly lower.

After the steel is heated uniformly to furnace temperature, it is withdrawn and placed in the high-heat furnace at the hardening temperature. When the steel has become heated uniformly at the hardening temperature, it is then quenched to harden it.

During the heating process, the metal absorbs heat, and its temperature rises until the point of decalescence is reached. At this point, additional heat is taken up by the metal. This heat is converted into work to change the pearlite into austenite without an increase in temperature until the process is completed. The total latent heat required to bring about these changes consists of internal latent heat and external latent heat. Both are utilized to fix permanently the structural change in the metal.

Quenching causes the metal to remain hard after it has been heated to the point of decalescence. If the metal were not quenched and allowed to cool slowly, the austenite would be reconverted to pearlite as the temperature decreases. This causes the metal to lose its hardness. When steel is cooled faster than its critical cooling rate, which is the purpose of quenching, a new structure is formed. The austenite is transformed into martensite. Martensite has an angular, needle-like structure and a very high hardness.

Martensite has a lower density than austenite. Therefore, the steel will increase in volume when quenched. Some of the austenite will not be transformed during quenching. The austenite will gradually change into martensite over a period of time. This change is known as ageing. This "ageing" results in an increased volume, which is objectionable in many items, such as gauges. A cold-treating process can be used to eliminate this problem.

Heating Baths

Other methods of heating are used in the hardening process. Various liquid baths are used with the following materials:

1. Lead
2. Cyanide of potassium
3. Barium chloride
4. Mixture of barium chloride and potassium chloride or other metallic salts

The lead bath cannot be used for high-speed steels because it begins to vaporize at 1190 degrees F (643.33 degrees C). The cyanide of potassium bath is used in gun shops extensively to secure ornamental effects and to harden certain parts.

Quenching or Cooling Baths

Although water is one of the poorest of all liquids for conducting heat, it is the most commonly used liquid for quenching metals in heat-treating operations. Water cools by means of evaporation.

To raise the temperature of 1 pound of water from 32 degrees F (0 degrees C) to 212 degrees F (100 degrees C) requires 180 Btu; to convert it into steam, an additional 950 Btu are required. Thus, the latent heat of vaporization (950 Btu) absorbed from the hot metal is the cooling agency. For efficient cooling, the film of steam must be replaced immediately by another layer of water—this requires circulation. If the heated metal is plunged into the water, thermocirculation results; however, if the metal is placed in the water horizontally, the film of steam that forms on the lower side of the piece is "pocketed," and the cooling action is greatly retarded. In still-bath quenching, a slow up-and-down movement of the work is recommended.

In addition to plain water, several other solutions can be used for quenching. Salt water, water and soap, mercury, carbonate of lime, wax, and tallow can also be used. Salt water will produce a harder scale. The quenching medium that will cool rapidly at the higher temperatures and more slowly at the lower temperatures should be selected. Oil quenches meet this requirement for many types of steel. There are various kinds of oils employed, depending on the nature of the steel being used.

The quenching bath should be kept at a uniform temperature so that successive pieces quenched will be subjected to the same conditions. The next requirement is to keep the bath agitated. The volume of work, the size of the tank, the method of circulation, and even the method of cooling the bath need to be considered in order to produce uniform results.

High-speed steel is usually quenched in oil. However, air is used for quenching many high-speed steels. Air under pressure is applied to the work by air blast.

Tempering

The purpose of tempering, or "drawing," is to reduce brittleness and remove internal strains caused by quenching. The process of tempering metal is accomplished by reheating steel that has been hardened previously and then quenching to toughen the metal and to make it less brittle. Unfortunately, the tempering process also softens the metal.

Tempering is a reheating process—the term "hardening" is often used erroneously for the tempering process. In the tempering process, the metal is heated to a much lower temperature than is required for the hardening process. Reheating to a temperature between 300 degrees F (149 degrees C) to 750 degrees F (399 degrees C) causes the martensite to change to troostue, a softer but tougher structure.

A reheating from 750 degrees F (149 degrees C) to 1290 degrees F (699 degrees C) causes a structure known as sorbite, which has less strength than troostite but much greater ductility. See Table 19.1 for tempering temperatures for various tools.

TABLE 19.1 Typical tempering temperatures for certain tools.

Degrees fahrenheit	Degrees celsius	Temper color	Tools
380	193	Very light yellow	Lathe center cutting tool for lathes and shapers
425	218	Light straw	Milling cutters, drill, and reamers
465	241	Dark straw	Taps, threading dies, punches, hacksaw blades
490	254	Yellowish Brown	Hammer faces, shear blades, rivet sets, and wood chisels
525	274	Purple	Center punches and scratch awls
545	285	Violet	Cold chisels, knives, and axes
590	310	Pale blue	Screwdrivers, wrenches, and hammers

Color Indications

Steel that is being heated becomes covered with a thin film of oxidation that grows thicker and changes in color as the temperature rises. This variation in color can be used as an indication of the temperature of the steel and the corresponding temper of the metal.

As the steel is heated, the film of oxides passes from a pale yellow color through brown to blue and purple colors. When the desired color appears, the steel is quenched in cold water or brine. The microscope can be used to explain the phenomena associated with the change in color. Steel consists of various manifestations of the same compound rather than separate compounds.

Although the color scale of temperatures has been used for many years, it gives only rough or approximate indications, which vary for different steels. The color scale for temper colors and corresponding temperatures is given in Table 19.2.

Specially prepared tempering baths equipped with thermometers provide a more accurate method of tempering metals. These are much more accurate than the color scale for tempering.

Casehardening

The casehardening operation is a localized process in which a hard "skin," or surface, is formed on the metal to a depth of 1/16 to 3/8 inch. This hard surface, or "case," requires two operations:

1. Carburizing where the outer surface is impregnated with sufficient carbon

2. Heat treating the carburized parts to obtain a hard outer case and at the same time give the "core" the required physical properties

TABLE 19.2 Temper colors of steel.

Colors	Temperatures (fahrenheit)	(Celsius)
Very pale yellow	430	221
Light yellow	440	227
Pale straw-yellow	450	232
Straw-yellow	460	238
Deep straw-yellow	470	243
Dark yellow	480	249
Yellow-brown	490	254
Brown-yellow	500	260
Spotted red-brown	510	266
Brown-purple	520	271
Light purple	530	277
Full purple	540	282
Dark purple	550	288
Full blue	560	293
Dark blue	570	299
Very dark blue	600	316

Carburizing is accomplished by heating the work to a temperature below its melting point in the presence of a material that liberates carbon at the temperature used. The material can be solid (charcoal, coke), liquid (sodium cyanide, other salt baths), or gas (methane, propane, butane). Often only part of the work is to be casehardened. There are four distinct ways by which casehardening can be eliminated from portions of the work:

1. Copper plating

2. Covering the portion that is not hardened with fire clay

3. Using a bushing or collar to cover the portion that is to remain soft

4. Packing with sand

Annealing

Annealing is a process of softening metal by heating it to a high temperature and then cooling it very slowly. The objective of the annealing processes is to remove stresses and strains set up in the metal by rolling or hammering so that it will be soft enough for machining.

The process of hardening is accomplished by increasing the temperature of a metal to the point of decalescence and then quenching it in a suitable cooling medium. The process of tempering is accomplished by reheating steel that has been hardened previously and then quenched to toughen the metal and make it less brittle. Tempering processes also can soften certain metals.

Hardfacing

Hardsurfacing, or hardfacing, is the melting of very hard metal powders and welding them in the surface of the base metal. There is a special type of welding rod used to hard-surface a tool. Some tools and machinery are subjected to much abuse. A farmer has rocks in his field, so a plow blade takes a lot of punishment and needs attention very often. The blade on a bulldozer also wears quickly and needs hardening. It can be protected against wear by hardening or coating the edge of the tool that receives the punishment. One method used to do this is to heat the forward area with a welding torch and then melti a special metal over it. The additional new metal will be extremely resistant to wear and will help prolong the life of the tool. One of the most common hardsurfacing materials are nonferrous cobalt-chromium-tungsten alloys. These include nickel-base, iron-base, and tungsten-carbide materials. A local welding-rod-supply dealer has information on each of the rods and where it can be used. The supply house usually has a catalog that lists all the rods and specialties.

Tool and Die Steels

Tool and die steels are classified as:

- High speed
- Hot work
- Cold work
- Shock-resisting
- Mold
- Special-purpose
- Water-hardening

Basic types of tool and die steels are shown below in Table 19.3.

In most instances, a good grade of tool steel is used for making punches and dies. The steel should be free from harmful impurities. Sometimes the body of the die can be made of cast iron with an inserted steel bushing to reduce the cost of material. An advantage of this type of construction is that the insert can be replaced when it becomes worn. Soft steel that has been case-hardened does not change its form as readily as tool steel, and any minute changes in form can be corrected readily, as the interior is soft.

Interior strains or stresses are set up in steel during the manufacturing process. In die-making operations, these stresses must be relieved before the die is brought to its final size, or they will cause distortion.

TABLE 19.3 Tool and die steels.

Type	AISI-SAE (grade)
High Speed	M (molybdenum base)
	T (tungsten base)
Hot work	H H 1 to H 19 (chromium base)
	H20 to H39 (tungsten base)
	H40 to H59 (molybdenum base)
Cold work	D (high carbon, high chromium)
	A (medium alloy, air hardening)
Shock resisting	S (oil hardening)
Mold steels	P1 to P19 (low carbon)
	P20 to P39 (others)
Special purpose	L (low alloy)
	F (carbon-tungsten)
Water hardening	W

MULTIPLE-CHOICE EXAM

1. Carbon steel that has had other metal added to it to improve its properties is called _____ steel.

    a. hard
    b. alloy
    c. soft
    d. none of these

2. Elements added to steel for alloying are:

    a. chromium
    b. molybdenum
    c. zirconium
    d. all of the above
    e. none of the above

3. Chromium, manganese, vanadium, tungsten, titanium, columbium, and molybdenum are _____ of deep hardening.

    a. carbides
    b. alloys
    c. carbons
    d. elements

4. At what percentage point is carbon-content steel the hardest possible?

    a. 0.40
    b. 0.60
    c. 0.20
    d. 0.10

5. Hardening metal is accomplished by increasing the temperature of a metal to its point of _____.

    a. irradiation
    b. decomposition
    c. decalescence
    d. incandescence

6. Steel with less than _____ percent carbon needs to be heated above its upper critical temperature.

    a. 0.65
    b. 0.75
    c. 0.95
    d. 0.85

7. The total latent heat required to bring about the changes needed for hardening consists of internal latent heat and _____ latent heat.

    a. external
    b. extensive
    c. extended
    d. none of the above

8. After the metal has been heated, _____is utilized to permanently fix the structural change in the metal.

    a. hammering   **b.** quenching

    c. freezing   **d.** none of the above

9. Martensite has a lower _____ than austenite.

    a. atomic weight   **b.** brittleness

    c. hardness   **d.** density

10. In the quenching process what is used to cool the steel by evaporation?

    a. oil   **b.** syrup

    c. water   **d.** chemical bath

11. High-speed steel is usually quenched in:

    a. oil   **b.** water

    c. chemicals   **d.** none of the above

12. The purpose of tempering is to reduce _____ and remove internal strains caused by quenching.

    a. scale   **b.** density

    c. hardness   **d.** brittleness

    e. ductility

13. Steel that is being heated becomes covered with a thin film of _____ that grows thicker and changes in color as the temperature rises.

    a. flux   **b.** rust

    c. scale   **d.** oxidation

14. Which is the distinct way to eliminate casehardening from portions of the work?

    a. copper plating

    b. covering the portion that is not hardened with fire clay

    c. using a bushing or collar to cover the portion that is to remain soft

    d. packing with sand

    e. none of the above

    f. all of the above

15. Tool and die steels are classified as _____.

    a. high-speed   **b.** hot work

    c. shock-resisting   **d.** mold steels

    e. all of the above   **f.** none of the above

TRUE-FALSE EXAM

1. Carbon steel that has had other types of metal added to it is called an alloy.

    True   False

2. Carbides are possible only when the base metal has sufficient carbon in the metal to form alloy carbides.

    True   False

3. High-carbon steel has a carbon content ranging from 0.60 to 1.5 percent.

    True   False

4. After the metal has been heated, quenching is utilized to fix permanently the structural change in the metal.

    True   False

5. Quenching causes the metal to remain hard after it has been heated to the point of decalescence.

    True   False

6. Martensite has a higher density than austenite.

    True   False

7. The lead bath cannot be used for high-speed steels because it begins to vaporize at 1190 degrees F.

    True   False

8. Water is the most commonly used liquid for quenching metals in heat-treating operations.

    True   False

9. The purpose of tempering metal is to increase brittleness and remove strain caused by quenching.

    True   False

10. The term "hardening" is often used erroneously for the tempering process.

    True   False

11. Steel that is being heated becomes covered with a thin film of oxidation.

    True   False

12. In determining tempering temperatures, the color brown-yellow denotes a temperature of 520 degrees F.

    True   False

13. In the casehardening process, a hard skin or surface is formed on the metal to a depth of 1/6 -to 3/8 inch.

True   False

14. Carburizing is accomplished by heating the work to a temperature above its melting point.

True   False

15. Hardsurfacing or hardfacing is the melting of very hard metal powders and welding them in the surface of the base metal.

True   False

MULTIPLE-CHOICE ANSWERS

1. B	4. B	7. A	10. C	13. D
2. D	5. C	8. B	11. A	14. F
3. A	6. D	9. D	12. D	15. E

TRUE-FALSE ANSWERS

1. T	5. T	9. F	13. T
2. T	6. F	10. T	14. F
3. T	7. T	11. T	15. T
4. T	8. T	12. T	

Chapter 20

SPECIALTY METALS: REACTIVE AND REFRACTORY

Specialty metals are those that are rather rare and serve special purposes in the electronics, space, and nuclear industries. They need special attention since they have different melting points and can cause fatalities if the welder is not aware of what he/she is working with at the time. Reactive and refractory metals were originally created to meet the needs of the nuclear and aerospace industries.

Reactive metals include:

- Beryllium
- Titanium
- Zirconium

Refractory metals include:

- Columbium
- Molybdenum
- Tantalum
- Tungsten
- Vanadium
- Zirconium

WELDING OF REACTIVE AND REFRACTORY METALS

Gas-tungsten-arc welding (GTAW), or TIG welding as it is commonly known, is the most widely used process for welding refractory alloy sheets up to 1/2 inch thick. Either argon or helium is used as the inert shielding gas. Never use carbon dioxide. Plasma-arc welding (PAW), braze-welding and brazing can also be used for thicknesses up to 3/4 inch. Following are the basic steps for welding these materials:

- Make sure there is a perfect fit-up as well as a thorough cleaning of the metal surface immediately prior to welding.

- Clean the surface by degreasing, followed by a steel brush and, if needed, sandblasting. Sandblasting is used to remove scale that is too difficult for a metal brush to remove. Stainless-steel brushes are recommended. Other types may leave rust or metal deposits on the surface.

- Do not expose the heated metal to the atmosphere. Protect it from atmospheric oxygen and other contaminants by shielding it in an inert-gas or vacuum chamber.

- Construct a dry box around the area to be welded and pump inert shielding gas into the space. The inert-gas chamber permits closer control of the atmospheric conditions around the weld.

- Ensure proper shielding when working outside a dry box. Use a trailing shield of inert gas to meet the contours of the weldment. An indication of inadequate shielding is a change in color of the electrode. The tip changes from bright to dull.

- Flux is not required when welding reactive and refractory metals.

- Use a filler metal similar in composition to that of the base metal with a similar thickness. Proper joint fit-up makes this unnecessary for thinner gauges.

- Weld slowly to avoid cracking. Preheating can reduce weld cracking. However, it must be done in a protective inert-gas atmosphere.

Beryllium and Beryllium Alloys

Beryllium is a lightweight reactive metal. It is characterized by extreme hardness (it can scratch a glass surface). Beryllium has a color similar to that of magnesium. It has a melting point of 2332 degrees F (1285 degrees C), a density comparable to that of magnesium, and a high electrical and thermal conductivity.

It is frequently used as an alloying element with other metals, such as copper, nickel and magnesium for increased strength, elasticity, and other characteristics. It is also used for its light weight, even lighter than aluminum, as well as for its tensile strength of about 55,000 psi.

Its relatively high melting temperature has suited it to applications in the aerospace industries. Other applications include the use of beryllium wire in the production of electrical circuits.

Note: Beryllium has a lower melting point when compared with the refractory metals tungsten, molybdenum, tantalum, and columbium. On the other hand, it does exhibit many of the same characteristics, especially in its need to be shielded from the contaminating effects of oxygen and other gases at high temperatures.

Beryllium is derived primarily from beryllium-aluminum silicates. The beryllium metal is prepared by electrolysis. Beryllium (1.0 to 2.5 percent) is combined with nickel (up to 1.0 percent) and copper to form the beryllium-copper alloy. This alloy has a relatively high tensile strength—up to 180,000 psi.

Beryllium is also used in small amounts (0.2-0.25 percent) as an alloy in the production of cast beryllium bronze. Iron, silicon, and cobalt are sometimes added to the beryllium-copper alloy to give it various desirable characteristics.

Caution: Beryllium dust and fumes are very toxic. Their extreme danger rests on the fact that there is no known cure (only remission agents) for their effects. In extreme cases breathing beryllium dust and fumes can be fatal. Every safety precaution should be taken when working with beryllium and beryllium alloys.

Beryllium can be joined by braze-welding and brazing. High-temperature welding processes are not recommended. Here are some specifics for welding beryllium:

- Use a low-temperature filler metal such as aluminum silicon or a silver-based alloy.

- Filler metal should be placed in the joint before braze-welding or brazing.

- Keep the braze-welding or brazing time as short as possible to prevent overheating the surface.

Columbium and Columbium Alloys

Columbium (or niobium) resembles steel in appearance. It is closely associated with tantalum in properties:

- Yellowish-white color

- Melting point of 4474 degrees F (2468 degrees C)

- Density slightly greater than iron

- Tensile strength of 48,000 to 59,000 psi in the annealed condition (up to 130,000 psi as drawn wire)

Columbium exhibits excellent strength, ductility, and corrosion-resistance characteristics. However, its resistance to corrosion is somewhat limited by its strong reaction to oxides. Oxidation becomes a serious problem at temperatures in excess of 752 degrees F (400 degrees C), necessitating the use of an inert shield. Columbium is a useful alloying element in other metals. It is particularly useful in imparting stability to stainless steel. Because of its low resistance to thermal neutrons, columbium is frequently used in nuclear equipment.

Columbium is added to steels in the form of ferrocolumbium (50.0 to 60.0 percent columbium, 33.0 to 43.0 percent iron, 7.0 percent silicon). Columbium (in amounts exceeding 75.0 percent) is alloyed with such metals as tungsten, zirconium, titanium, and molybdenum to produce columbium alloys that will retain high tensile strength at high temperature (in the 2000- to 5000-degree F range). Columbium is a refractory metal, a characteristic that encourages frequent use of columbium alloys in the production of high-temperature-resisting components for missiles, turbines, and jet engines.

Gas-tungsten-arc welding (GTAW) in a vacuum chamber is the method commonly used to join columbium and the columbium alloys. Braze-welding and brazing can also be used to join columbium. Here are some tips for welding with these metals:

- Preheating is required for some columbium alloys to avoid weld cracks.

- Heat treatment should be performed in an inert-shielding-gas or vacuum chamber.

- A fast welding speed with minimum amp input produces the best penetration.

- Copper backing bars should be used to extract heat from the weld.

Molybdenum and Molybdenum Alloys

Molybdenum resembles steel in color and tungsten in most of its physical properties. It has a melting point of 4730 degrees F (2610 degrees C). As such, it comprises one of the refractory metals (those having a melting point above 3600 degrees F) and can be grouped with columbium (niobium), tungsten, and similar metals. Molybdenum is a very important alloying element in the production of iron and steel. It is not generally used in quantities exceeding 4 percent. Research in the development of molybdenum steels began in the 1890s, but it was not until after World War I that it was possible to produce this type of steel on an economical basis. It received widespread use in the automotive industry and rapidly expanded the extent of its application to other areas. Molybdenum is now widely used in the production of tool steels and high-speed steels, where it contributes to wear resistance, hardness, and strength.

Molybdenum is also added to gray iron in amounts of less than 1.25 percent to increase its tensile strength and hardenability. Molybdenum is added to steel to:

- Increase hardness

- Increase endurance

- Increase corrosion resistance (stainless steel)

- Increase the tendency toward deep hardening

Due to the high melting point of molybdenum and its alloys, they are employed in the manufacturing of rocket and gas-turbine engines. Its greatest use is in the electronics and nuclear industries.

Gas-tungsten-arc-welding (GTAW) and gas-metal-arc-welding (GMAW) processes, both with inert shielding gases, are used to weld molybdenum and its alloys. Brazing also can be used to join the thinner gauges of molybdenum. Here are some guidelines for welding with these metals:

- Use argon or helium as the inert shielding gas. Never use carbon dioxide as a shielding gas when welding molybdenum.

- Preheating is usually not necessary.

- When brazing be sure to select a brazing-filler metal suitable for the application. Temperatures for brazing filler metals range from 1200 to 4500 degrees F.

- Keep the brazing time as short as possible to minimize heat buildup.

Tantalum and Tantalum Alloys

Tantalum is a refractory metal in the same category as tungsten, molybdenum, and columbium. It has a color ranging from whitish to silvery gray, a tensile strength of 50,000 psi, and a melting point of 5425 degrees F. It is characterized by an unusually strong resistance to acids and corrosion. For this reason it finds useful application in the manufacture of chemical and surgical equipment. Tantalum and its alloys have many other interesting applications, including its use as filaments in light bulbs and as components in jet engines and rocket motors. Because tantalum is a very ductile metal, it can be drawn or rolled without annealing.

Tantalum has an extremely strong reaction to the oxygen in the surrounding atmosphere or in the oxygen cutting stream (oxyacetylene cutting torch). For this reason, tantalum loses much of its effectiveness as a high-temperature structural metal. Commercially pure tantalum is soft and ductile. A thin oxide surface layer similar to the one found on aluminum covers the surface of tantalum. It is this oxide surface layer that contributes to the resistance of tantalum to acids.

Gas-tungsten-arc welding (GTAW) is the most commonly used welding process for joining tantalum and its alloys. Brazing is also used in certain applications. Here are some tips:

- Shield the weld metal and surrounding area with an appropriate inert gas.

- Use copper brazing in an inert gas vacuum to join thin gauges of tantalum.

- Use the same welding processes as for columbium with tantalum, since the characteristics are similar.

Titanium and Titanium Alloys

Titanium has a silvery color, a high resistance to corrosion, and the highest affinity for carbon of all the known metals. As a result of this carbon affinity, titanium functions as an alloying element (less than 1 percent is needed) to stabilize the carbon in steels and prevent cracking. Another characteristic of titanium is its strong tendency to form carbides. For this reason it is used in some chromium steels to counteract chromium-depletion tendencies.

Titanium has a melting point of 3035 grees F (1800 degrees C). The titanium alloys are as strong as steel, but are 50 to 60 percent lighter in weight. Commercially pure titanium has a tensile strength of 45,000 psi, and the hardened titanium alloys may have tensile strengths extending to 200,000 psi. The ductility of this metal is relatively low.

Cleaning before welding titanium. Perfect fit-up as well as a thorough cleaning of the metal surface is required prior to welding. Cleaning the surface may be done by degreasing, followed with a steel brush and sandblasting. Stainless-steel brushes are recommended because other types may leave rust or metal deposits on the surface. Sandblasting is used to remove scale that is too difficult for the metal brush. All cleaning must be done immediately prior to welding. The surface must be thoroughly cleaned of all contaminants.

Welding titanium. Due to its strong attraction as an absorbent for most solids and gases (except argon and helium), it must be welded under specially shielded conditions. In addition to the use of argon or helium as the shielding gas in gas-tungsten-arc welding, a trailing shield should also be used until the surface has cooled to approximately 750 degrees F.

The gas-tungsten-arc (GTAW), gas-metal-arc (GMAW), plasma-arc (PAW), and braze, welding processes, as well as brazing, can be used to join all thicknesses of titanium and titanium alloys. Oxyacetylene welding is not recommended for welding this metal and its alloys.

Here are some tips:

- In addition to the use of argon or helium as the shielding gas in the gas-tungsten-arc welding (GTAW) process, a trailing shield should also be used until the surface has cooled to approximately 750 degrees F.

- Filler metals similar in composition to that of the base metal are used with certain thicknesses of titanium. Proper joint fit-up makes this unnecessary for thinner gauges of titanium and titanium alloys.

- Flux is not required.

Tungsten and Tungsten Alloys

The color of tungsten ranges from steel gray to a silvery white. It has a melting point of 6039 degrees F (3337 degrees C), the highest of all metals. As a result of this characteristic, tungsten has found applications in the aerospace industry for rocket components that must withstand extreme temperatures. Tungsten is also used in the electronics industry, in the electrical industry, and in the production of high-speed cutting tools.

Tungsten is an important alloy in the production of steels. Important among these steels are the tungsten tool steels (up to 20.0 percent tungsten), and particularly the high-speed tool steels. As an alloying element, tungsten contributes to steel such characteristics as hardness, strength, and wear resistance. Here are some tips for welding tungsten:

- Tungsten and tungsten alloys have weldability characteristics very similar to those described for molybdenum and its alloys.

- Torch brazing can be used to join tungsten and its alloys.

- Do not use a nickel-base filler metal when brazing tungsten. The interaction between nickel and tungsten can cause weld problems.

Vanadium and Vanadium Alloys

Vanadium is a grayish-white metal used primarily in steel production. It is also used as an alloying element for other metals. Vanadium tends to increase the hardenability of a metal and to reduce or eliminate the harmful effects of overheating.

Vanadium melts at a temperature of 3236 degrees F. Because it is not subject to oxidation (it is highly resistant to corrosion), it serves as a strong deoxidizer in steels. It increases tensile strength, has little or no effect on ductility, and greatly increases grain growth. In the annealed condition vanadium has a tensile strength of 66,000 psi. Some steels in which vanadium is present as a major alloying element are:

- Python steel (0.25 percent vanadium)

- Vasco vanadium (0.20 percent)

- SAE 6145 steel (0.18 percent)

- Colonial No. 7, Elvandi, and so forth

These metals are known in general as vanadium steels. This general grouping is frequently divided into subgroups such as chromium-vanadium steels or, depending on the alloys present.

Note: Use the same welding tips listed for zirconium and the zirconium alloys to weld vanadium and its alloys. The weldability of both metals is very similar.

Zirconium and Zirconium Alloys

Zirconium is a silvery-white metal with a melting point of 3330 degrees F (1700 degrees C) or almost three times that of aluminum. It has metallurgical characteristics very similar to those of titanium. Pure zirconium has a relatively low tensile strength of approximately 32,000 psi. It exhibits strong corrosion resistance and excellent thermal stability. A principal application of zirconium is as an absorber of gases in electronic tubes.

It is also used:

- As an alloying element in steel to prevent brittleness and age hardening

- In zirconium sheets

- In zirconium foil

- In zirconium wire

- In zirconium rods

- In surgical tools and components

- In chemical equipment

Its thermal stability and resistance to corrosion make it highly suited for construction in the nuclear field. Zircaloy-2™ and Zircaloy-3™ are two of the better-known trade names for zirconium alloys. Zircaloy-2, a product of Westinghouse Electric Corporation, contains 98.28 percent zirconium. The other elements (tin, iron, chromium, nickel) are added to increase its strength and to reduce the tendency of zirconium to absorb hydrogen. Both Zircaloy-2 and Zircaloy-3 have excellent weldability.

Gas-tungsten-arc welding (GTAW) or gas-metal-arc welding (GMAW) are the methods commonly used to join zirconium and the zirconium alloys. Following are some tips:

- Zirconium reacts to both oxygen and nitrogen, which makes welding in an inert-gas-shielding atmosphere necessary.

- Critical welds should be performed inside a dry box.

- Maintain shielding until the weld metal and the heat-affected zone have cooled to at least 700 degrees F.

- The rate of cooling can be safely increased by using copper backup plates and clamps.

- GMAW-welding wire must be of the hard-drawn type.

Other Specialty Metals

Several uncommon metals have become important in modern technology as a result of their unique properties. Hafnium, thorium, and beryllium, for instance, are used in nuclear reactors because of their low neutron-absorption characteristics.

High-temperature applications often require metals with high melting points. These include niobium (also known as columbium), molybdenum, tantalum, rhenium, and tungsten. Depleated uranium, because of its very high density, is useful in special applications. It is used where maximum weight must be put into a limited space, as in counterweights. Zirconium is used for its outstanding corrosion resistance to most acids, chlorides, and organic acids. When alloyed with a small percentage of hafnium, it has a yield strength of about 579 MPa (84,000 psi) and a tensile strength of about 620 MPa (9000) psi.

With the high interest in space, nuclear, high-temperature and electronic applications, a considerable amount of research and development work is being done on these uncommon metals.

MULTIPLE-CHOICE EXAM

1. Reactive metals include beryllium, titanium, and _____.
    a. columbium
    b. tungsten
    c. tantalum
    d. zirconium

2. Refractory metals include molybdenum, vanadium, tungsten, columbium, zirconium, and _____.
    a. lead
    b. zinc
    c. tantalum
    d. beryllium

3. Which welding process is the most widely used for welding refractory alloy sheets up to 1/8 inch thick?
    a. arc
    b. TIG
    c. MIG
    d. shield-metal-arc

4. Which welding process is the most commonly used for welding refractory alloy sheets up to 3/4 inch thick?
    a. GTAW
    b. plasma-arc (PAW)
    c. submerged-arc
    d. GMAW

5. Flux is _____ required when welding reactive and refractory metals.
    a. always
    b. not usually
    c. usually
    d. not

6. Beryllium is a lightweight _____ metal.
    a. reflective
    b. refractory
    c. reactive
    d. alloy

7. Beryllium dust and fumes can be _____.
    a. fatal
    b. nauseating
    c. crippling
    d. none of the above

8. Beryllium can be joined by both braze welding and _____.
    a. submerged welding
    b. brazing
    c. plasma welding
    d. arc welding

9. Columbium resembles _____ in appearance.
    a. iron
    b. gray iron
    c. cast iron
    d. steel

10. When alloyed with stainless steel, columbium can add _____ to the mixture.
    a. stability
    b. oxidation
    c. weight
    d. none of the above

11. Which type of welding is needed to join columbium and columbium alloys?
    a. gas-tungsten-arc
    b. submerged-arc
    c. plasma-arc
    d. none of the above

12. Preheating is required for some columbium alloys to avoid weld _____.
    a. cracks
    b. peeling
    c. breaks
    d. weakness

13. Molybdenum is added to gray iron to increase its tensile strength and _____.
    a. hardenability
    b. melting temperature
    c. brittleness
    d. ductility

14. When do you use carbon dioxide as a shielding gas while welding molybdenum?
    a. occasionally
    b. always
    c. never
    d. none of the above

15. Tantalum is a _____ metal that belongs to the same category as tungsten, columbium, and molybdenum.
    a. refractory
    b. reactive
    c. hard
    d. soft

16. Which is he most commonly used welding process for tantalum and its alloys?
    a. GMAW
    b. GTAW
    c. oxyacetylene
    d. plasma-arc

17. Titanium has a silvery color, a high resistance to corrosion, and the highest affinity for _____ of all the known metals.
    a. oxygen
    b. carbon
    c. nitrogen
    d. helium

18.   Titanium has an advantage over steel inasmuch as it is just as strong but much _____.

    a. more ductile   **b.** lighter

    c. stronger   **d.** none of the above

19.   Which welding methods can be used to weld titanium and titanium alloys?

    a. GTAW   **b.** GMAW

    c. PAW   **d.** braze

    e. all of the above   **f.** none of the above

20.   Use the same welding tips listed for zirconium and the zirconium alloys to weld _____ and its alloys.

    a. titanium   **b.** zirconium

    c. tantalum   **d.** vanadium

TRUE-FALSE EXAM

1.   Reactive and refractory metals were originally created to meet the needs of the nuclear and aerospace industries.

    True   False

2.   Reactive metals include beryllium and titanium.

    True   False

3.   Refractory metals include columbium and vanadium.

    True   False

4.   GTAW welding is used for welding refractory alloys.

    True   False

5.   Never use carbon dioxide as a shielding gas in welding reactive and refractory metals.

    True   False

6.   Beryllium is a heavy reactive metal.

    True   False

7. Beryllium dust and fumes are very toxic.

True   False

8. Columbium resembles steel in appearance.

True   False

9. Columbium is used primarily for making alloys that are used in high temperature-resisting components such as missiles, turbines, and jet engines.

True   False

10. Preheating is required for some columbium alloys to avoid cracks.

True   False

11. Molybdenum is a very important alloying element in the production of iron and steel.

True   False

12. Molybdenum is added to steel to increase its hardness, endurance, and corrosion resistance.

True   False

13. Due to its low melting point, molybdenum is used in the manufacturing of rock- and gas-turbine engines.

True   False

14. Never use carbon dioxide as a shielding gas when welding molybdenum.

True   False

15. Tantalum has an extremely strong reaction to the nitrogen in the surrounding atmosphere or to the oxygen cutting stream used in gas welding with acetylene.

True   False

16. Titanium is a very lightweight metal.

True   False

17. Titanium must be welded under specially shielded conditions.

True   False

18. Oxyacetylene welding is highly recommended for welding titanium.

True   False

19. Tungsten and the tungsten alloys have weldability characteristics very similar to those described for molybdenum and its alloys.

    True   False

20. Do not use a nickel-base filler metal when brazing tungsten.

    True   False

MULTIPLE-CHOICE ANSWERS

1. D	5. D	9. D	13. A	17. B
2. C	6. C	10. A	14. D	18. B
3. B	7. A	11. A	15. A	19. E
4. B	8. B	12. A	16. B	20. D

TRUE-FALSE ANSWERS

1. T	6. F	11. T	16. T
2. T	7. T	12. T	17. T
3. T	8. T	13. F	18. F
4. T	9. T	14. T	19. T
5. T	10. T	15. F	20. T

Chapter 21

WELDING GALVANIZED METALS

Welding Licensing Exam Study Guide

Metal parts are given corrosion-resistant coatings by being dipped into certain molten metals. The most commonly used are zinc, tin, and an alloy of lead and tin.

Hot-dipped galvanizing is the most widely used method of providing steel with a protective coating. After the parts, or sheets, have been cleaned, they are fluxed by dipping them into a solution of zinc chloride and hydrochloric acid. They then are dipped into a molten zinc bath. The resulting zinc coating is complex, consisting of a layer of $FeZn_2$ at the metal surface, an intermediate layer of $FeZn_7$, and an outer layer of pure zinc. Hot-dip galvanizing provides a good degree of corrosion resistance.

The coating thickness should be controlled since coatings that are too thick crack and peel. A wide variety of spangle patterns can be obtained by proper processing. When galvanizing is done properly, considerable subsequent bending and forming can be done without damaging the coating. However, rimmed steel should not be galvanized.

Galvanized iron and steel are ferrous metals coated with a thin layer of zinc for protection against corrosion. Galvanized steel is widely used in the building-construction industry for siding, roofs, downspouts, and gutters.

Any of the fusion-welding processes can be used to join galvanized iron or steel. Shielded-metal-arc welding (SMAW) is probably the most widely used process with galvanized metals.

Caution: The heat of the arc or flame liberates zinc fumes during welding. Provision must be made to protect the welder from these fumes by providing adequate ventilation if the welding is being done inside and wearing the necessary respirator protection. This precaution should be taken at all times and for every job involving galvanized metals. If the welder should experience any nausea during or after welding galvanized material, medical help should be sought immediately.

Here are some tips for welding galvanized metals:

- The American Welding Society advises removing the zinc coating on galvanized steels before welding, brazing, or soldering. Any heat applied to the zinc coating will release potentially harmful zinc fumes into the atmosphere. Remove 2 to 4 inches of the zinc coating on both sides of the intended weld line and on both sides of the work piece. Grinding away the zinc coating is the preferred method.

- Burning the zinc coating away from the weld zone is sometimes used instead of grinding the surface.

- Removing the zinc coating from the weld zone is not an absolute requirement. Check the job specifications for guidance.

TABLE 21.1 Recommendations for welding/joining galvanized metals.

Welding or joining process	Comments
Shielded Metal-Arc Welding (SMAW)	• Use flux-covered electrodes when welding galvanized metals with the SMAW process. • Reduce the electrode angle by about 30° and whip it back and forth to push the molten zinc pool away from the weld. Note: the electrode angle and whipping motion results in a reduced welding speed. • Use a wider than normal root opening to facilitate penetration. • Remove the spatter after the weld is finished. Slightly more spatter is produced when using the SMAW arc to remove the zinc coating than is the case when welding an uncoated surface.
Gas Metal-Arc Welding (GMAW)	• Commonly recommended for welding thinner gauges of galvanized metals. • The weld mechanical properties are unaffected by the zinc coating. • The GMAW process requires a higher heat input and lower welding speeds for zinc coated materials. Both are required to burn off and remove the coating. • There may be a reduction in welding speed if the zinc has not been removed because the galvanized coating must be burned off ahead of the weld. • Use a 100% CO_2 gas to shield the weld. • The arc is stable, but penetration is less than for uncoated metals.
Oxyacetylene Welding (OAW)	• Use the same procedure recommended for welding uncoated steel. • Move the filler rod back and forth to produce a ripple weld. • Because low travel speed is necessary to bring the joint edges to the fusion temperature, the extra heat causes the zinc coating appearance to be affected over a much greater area than other welding processes.
Braze Welding	• Braze welding (as well as brazing) is used where full strength welds are not required, (where a color match between the weld material and the base metal is not necessary, or where high temperatures cannot be used. • No preheating is required when braze welding steel. • Use deoxidizing flux when braze welding galvanized meals. A deoxidizing flux prevents atmospheric oxygen from attacking the zinc coating while it is in the molten state. • When braze welding galvanized pipe, be sure to coat both the inside and outside of the pipe with the liquid flux about 3 inches back from the joint. When the deposited metal cools, the contraction caused by the cooling draws the ends of the pipe together.
Brazing	• Coat both sides of the galvanized sheets along the seam with a thick coating of a paste flux. The paste coating protects the galvanized coating • Braze with a low temperature brazing rod or a nickel brazing rod together with a matching flux. Do not use a bronze rod with a higher temperature melting point. A higher temperature rod will burn the galvanized coating.
Soldering	• Used for joining furnace ductwork, gutters and downspouts. • Electro-galvanized steel sheet is easier to solder than steel sheet covered with hot dipped zinc coatings. • Chromate treatment of zinc coated sheets may interfere with solder flow.

MULTIPLE-CHOICE EXAM

1. Metals parts are given corrosion-resistant coatings by being dipped into certain metals. These metals are:

    a. zinc   **b.** lead and tin alloy

    c. tin   **d.** all of the above

    e. none of the above

2. Hotdipped galvanizing is the most widely used method of providing steel with a _____ coating.

    a. protective   **b.** solid

    c. tin   **d.** lead

3. The coating thickness should be controlled, since coatings that are too thick will crack and _____.

    a. turn gray   **b.** peel

    c. discolor   **d.** none of the above

4. Galvanized _____ and steel are ferrous metals coated with a thin layer of zinc for [run in] protection against corrosion.

    a. iron   **b.** aluminum

    c. molybdenum   **d.** none of the above

5. Any of the _____ welding processes can be used to join galvanized iron or steel.

    a. nonfusion   **b.** soldering

    c. fusion   **d.** brazing

6. What is the most widely used welding process for welding galvanized metals?

    a. SMUP   **b.** SMPW

    c. SMAP   **d.** SMAW

7. How do you make sure that the galvanized-steel welding process does not create a health hazard for the welder?

    a. weld outdoors   **b.** provide good ventilation

    c. wear a respirator   **d.** all of the above

    e. none of the above

8. Two methods of removing the galvanizing coating from steel are burning and _____.

    a. overheating the area   **b.** using paint remover

    c. grinding   **d.** none of the above

9. GMAW is recommended for welding of _____ gauges of galvanized steel.

    a. thinner   **b.** thicker

    c. larger   **d.** none of the above

10. Electro-galvanized steel sheet is easier to _____ than steel covered with hot-dipped zinc coatings.

    a. solder   **b.** braze

    c. weld   **d.** none of the above

TRUE-FALSE EXAM

1. The commonly used metals for coatings on steel and iron are zinc, tin, and an alloy of lead and tin.

True   False

2. Hydrochloric acid is used to flux hot-dipped metals.

True   False

3. Hot-dipped galvanizing provides a good degree of corrosion resistance.

True   False

4. The AWS advises removing the zinc coating on galvanized steels before welding, brazing, or soldering.

True   False

5. Burning the zinc coating from the weld zone is sometimes done instead of grinding the surface.

True   False

6. Removing the zinc coating from the weld zone is not an absolute requirement.

True   False

7. Use flux-coated electrodes when welding galvanized metals with the SMAW process.

True   False

8. The GMAW process requires a higher heat input and lower welding speeds for zinc-coated materials.

    True   False

9. Braze welding (as well as brazing) is used where full-strength welds are required.

    True   False

10. When brazing galvanized sheet steel, coat both sides of the galvanized sheets along the seam with a thick coating of boric acid.

    True   False

MULTIPLE-CHOICE ANSWERS

1. D	3. B	5. C	7. D	9. A
2. A	4. A	6. D	8. C	10. A

TRUE-FALSE ANSWERS

1. T	4. T	7. T	9. F
2. T	5. T	8. T	10. F
3. T	6. T		

—NOTES—

Part G

ALLOYS

—NOTES—

Chapter 22

WELDING SOFT METALS AND THEIR ALLOYS

A lloys extend the uses of metals in many ways. Steel is alloyed to produce desirable characteristics for a specific application. Aluminum, another favorite of the construction and transportation industries as well as aerospace, is a light-weight metal with a number of alloy mixes to fit a specific job.

ALUMINUM

Aluminum has one-third the weight of steel. Aluminum alloy 7178-T6 is heat-treatable and can develop strengths up to 88,000 psi. With this strength-to-weight ratio, aluminum has an advantage over steels especially in aerospace applications.

Aluminum has excellent forming characteristics. It can be bent, extruded, joined, or machined with comparative ease. Aluminum is an excellent conductor of electricity and makes a good heat-conductor material. There is the added advantage of having high atmospheric corrosion resistance.

Many alloys of aluminum have been developed by metallurgists for special products and other applications. Aluminum alloys includee:

- Aluminum copper
- Aluminum manganese
- Aluminum silicon
- Aluminum magnesium
- Aluminum silicon magnesium
- Aluminum zinc

Wrought-Aluminum Alloys

A classification system for wrought-aluminum alloys was established by the Aluminum Association in October 1954. The standardized system consisted of a four-digit number with these specifications (see Table 22.1):

- First number identifies the alloy group
- Second number identifies the impurity limit in the original alloy
- Last two digits identify either the specific alloy or the purity of the aluminum

TABLE 22.1 Major alloying elements.

Aluminum, 99.00% and greater	1xxx
Copper	2xxx
Manganese	3xxx
Silicon	4xxx
Magnesium	5xxx
Magnesium and silicon	6xxx
Zinc	7xxx
Other element	8xxx

TABLE 22.2 Major alloying elements for castings.

Aluminum, 99.00% and greater	1xx.x
Copper	2xx.x
Silicon with copper and/or Mg	3xx.x
Silicon	4xx.x
Magnesium	5xx.x
Magnesium and silicon	6xx.x
Zinc	7xx.x
Tin	8xx.x
Other elements	9xx.x

With aluminum alloy 7178-T6, for example, the first digit (7) indicates that the alloy is of the zinc group. The second digit (1) indicates the original alloy modifications; numbers 1 through 9 can be used to show alloy modifications or impurity limits of the alloy group. The last two numbers (78) express the minimum aluminum percentage to the nearest 0.01 percent. The major alloying element is aluminum, thus the 7178 alloy has the purity of 99.78 percent aluminum with a level of control over alloy modification. Zinc is the alloy group being used. A dash separates the numbers from the letter T and indicates the condition of the alloy. It can withstand 88,000 pounds per square inch (psi) and was solution-treated and artificially aged. The metal used a specific heat treatment to provide this exceptional strength. (See Table 22.2.)

Its low melting temperature tends to make it suitable for casting. However, pure aluminum is seldom cast. Its high shrinkage and susceptibility to hot cracking cause considerable difficulty, and scrap is high. By adding small amounts of alloying elements it is very suitable for casting. Its characteristics are improved and strength is increased. Large amounts of aluminum alloys are cast; the principal alloying elements are copper, silicon, and zinc.

Welding Aluminum

Aluminum can be welded with oxyacetylene, but it is a tricky job because the metal has a very low melting point and gives no warning when it is being reached.

Begin by preheating the metal with an excess acetylene flame. This flame will deposit carbon (soot) on the surface and make it look very black. Next, heat the joint area with a neutral flame until the carbon is removed. Then apply some special liquid-aluminum welding flux.

While keeping the cone of the flame at least 1 inch above the joint, use a special aluminum welding rod in the flame to tin the joint. After the joint has been tinned, melt more rod onto the area and build up the joint.

Once the buildup has been completed, be sure to keep the flame of the torch away from the joint, since this may cause the aluminum to become very weak. After the weld has been completed, be sure to remove all flux from the joint and surrounding area. Use warm water and a brush.

Zinc Alloys

Aluminum alloyed with zinc produces an aluminum alloy with a low melting point. Pure zinc is a metal with a low melting point. It can easily be cast at temperatures ranging from 750 to 800 degrees F. Because zinc can be cast under great pressure at these low temperatures, it produces a very accurate and consistent product.

The machinability of aluminum-base alloys varies greatly. Most cast alloys are machined easily. For wrought alloys, with the exception of a few special types, special tools and techniques are desirable if large-scale machining is to be done.

Filler Materials

A number of variables need to be considered during the selection of the most suitable filler alloy for a particular base alloy. When choosing the optimum filler alloy, both the base alloy and the desired performance of the weld must be considered in terms of what it is subjected to and what it is expected to do.

Aluminum alloys along with their various tempers make up a wide and versatile range of manufacturing materials. For most product designs it is important to understand the differences among the many alloys and their performance and weldability characteristics. For instance, when developing arc-welding procedures for different alloys, keep in mind the specific alloy being welded.

Soldering

Chapter 12 covers soldering of metals such as aluminum. Fluxes, temperatures, and procedures are covered in detail.

COPPER AND COPPER ALLOYS

Copper is rarely used in its pure state except in the electrical industry. Copper and copper alloys are some of the most important engineering materials. This is due to their characteristics:

1. Good electrical and thermal conductivity

2. Good corrosion resistance

4. Good metal-to-metal wear resistance

5. Distinct aesthetic appearance

6. Can be joined by welding, brazing, and soldering

Preparing the Surface

The weld area should be clean and free of oil, grease, dirt, paint, and oxides prior to welding. Use a wire brush with a bronze wire to do the brushing. Follow up by degreasing with a suitable cleaning agent.

The oxide film formed during welding should also be removed with a wire brush after each weld run is deposited.

Preheating

Welding of thick copper sections requires a high preheat. This is due to the rapid conduction of heat from the weld joint into the surrounding base metal.

Most copper alloys, even in thick sections, do not require preheating because the thermal diffusivity is much lower than for copper. To select the correct preheat for a given application, consider the welding process. The alloy being welded, the base-metal thickness, and the overall mass of the weld should be considered.

Aluminum-bronze and copper-nickel alloys should not be preheated. It is desirable to limit the heat to as localized an area as possible to avoid bringing too much of the material into a temperature range that will cause a loss in ductility. It is also important to be sure the preheat temperature is maintained until welding of the joint is completed.

Zinc is the most often used alloying material for copper. Most of these alloys are known as brass.

TABLE 22.3 Standard designations for copper and copper alloys.

	Wrought Alloys		Cast Alloys
100–155	Commercial coppers	833–838	Red brasses and leaded red brasses
162–199	High-copper alloys		
200–299	Copper–zinc alloys (brasses)	842–848	Semi-red brasses and leaded semi-red brasses
300–399	Copper–zinc–lead alloys (leaded brasses)		
		852–858	Yellow brasses and leaded yellow brasses
400–499	Copper–zinc–tin alloys (tin brasses)		
		861–868	Manganese and leaded manganese bronzes
500–529	Copper–tin alloys (phosphor bronzes)		
		872–879	Silicon bronzes and silicon brasses
532–548	Copper–tin–lead alloys (leaded phosphor bronzes)		
		902–917	Tin bronzes
600–642	Copper–aluminum alloys (aluminum bronzes)	922–929	Leaded tin bronzes
		932–945	High-leaded tin bronzes
647–661	Copper–silicon alloys (silicon bronzes)	947–949	Nickel–tin bronzes
		952–958	Aluminum bronzes
667–699	Miscellaneous copper–zinc alloys	962–966	Copper–nickels
		973–978	Leaded nickel bronzes
700–725	Copper–nickel alloys		
732–799	Copper–nickel–zinc alloys (nickel silvers)		

Gas-Metal-Arc Welding of Copper and Copper Alloys

Copper and copper alloys are hardened by cold working and, in commercial grades, are available in various degrees of hardness.

ERCu copper electrodes are recommended for GMAW with copper. Deoxidized copper is a versatile 98-percent-pure copper alloy. The gas mixture used will be largely determined by the thickness of the copper section to be welded. Argon is generally used for thicknesses of 6 mm and under.

Helium-argon mixtures are used for welding thicker sections. The filler metal should be deposited with stringer beads or narrow-weave beads using spray transfer. Table 22.3 above gives standard designations for copper and copper alloys.

Gas-Metal-Arc Welding of Copper-Tin and Phosphor-Bronze Alloys

These alloys have a wide solidification range that gives a coarse grain structure. Care must be taken during welding to prevent cracking of the weld metal. Hot-peening of the weld metal will reduce the stresses developed during welding and the likelihood of cracking. The weld pool should be kept small, using stringer beads at high travel speed.

Gas-Tungsten-Arc Welding of Copper and Copper Alloys

Copper sections up to 16.0 mm in thickness can be successfully welded using the gas-tungsten-arc welding (GTAW) process. The recommended filler wire is a filler metal whose composition is similar to that of GTAW of copper and copper alloy for sections up to 1.6 mm thick.

For sections over 1.6 mm thick argon shielding gas or helium mixes are preferred. In comparison to argon, argon/helium mixes permit deeper penetration and higher travel speeds at the same welding current. A 75-percent helium/25-percent argon mixture is commonly used for good penetration combined with the easy arc starting and improved arc stability.

Forehand welding is preferred, with stringer beads or narrow weave beads.

Recommended shielding gases include welding-grade argon combined with up to 3 percent oxygen or 25 percent helium or helium combined with 25 percent argon.

Gas-Metal-Arc Welding of Copper-Silicon Alloys

ERCuSi-A-type welding consumables plus argon shielding and relatively high travel speeds are used with this process.

Silicon bronze is a copper-based wire recommended for GMAW of copper-silicon alloys. It is important to ensure that the oxide layer is removed by wire brushing between passes. Preheating is unnecessary, and interpass temperature should not exceed 212 degrees F (100 degrees C).

Brazing Copper and Copper Alloys

The principle of brazing is to join two metals by fusing with a filler metal. The filler metal must have a lower melting point than the base metals but greater than 842 degrees F (450 degrees C); use of a filler metal with a melting point less than this is soldering. The filler metal is usually required to flow into a narrow gap between the metals by capillary action.

Brazing is used widely for joining copper and copper alloys, with the exception of aluminum bronzes containing greater than 10 percent aluminum and alloys containing greater than 3 percent lead.

Establish the correct heating pattern so that the filler metal flows up the thermal gradient into the joint.

To prepare the surface, standard solvent or alkaline degreasing procedures are useful in the cleaning of copper-base metals. Be careful when using mechanical methods to remove surface oxides. Chemical removal of surface oxides can be accomplished by using an appropriate pickling solution such as Chrome Bright®.

Control the distance between the joints to be joined. They must be controlled to within certain tolerances. The tolerances depend upon the brazing alloy and the parent metal used. The best joint gap typically lies between 0.04 and 0.20 mm.

To achieve an adequate bond during brazing, the following points should be considered:

- Joint surfaces must be clean and free of oxides, and provision must be made for the correct joint gap for the particular brazing filler.

- Usually a joint overlap of three or four times the thickness of the thinnest member to be joined will work.

- The goal is to use as little material as possible to achieve the desired strength.

Figure 22.1 illustrates common joint design for silver brazing.

Use a neutral flame (one in which equal amounts of oxygen and acetylene are mixed at the same rate). The white inner cone of the flame is clearly defined and shows no haze.

Flux residue must be removed. Use one of the following methods:

- A dilute solution of hot caustic-soda dip

- Wire brushing and rinsing with hot water

- Wire brushing and steam

Incomplete flux removal may cause weakness and failure of the joint.

Braze Welding of Copper

Braze and fusion welding are similar except that the filler metal uses a lower melting point than the parent metal. The braze-welding process gets its strength from the tensile strength of the filler metal deposited and the bond strength developed between the filler metal and the parent metal. Oxyacetylene is usually preferred because of its easier flame setting and rapid heat input.

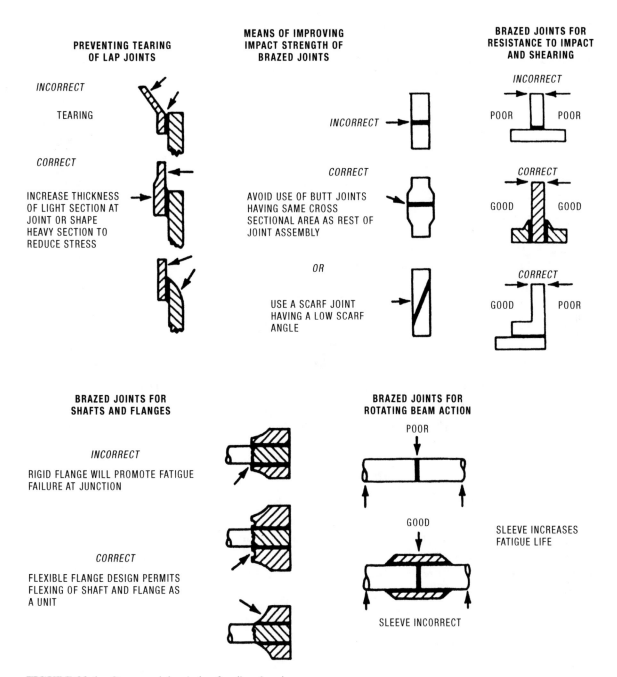

FIGURE 22.1 Common joint design for silver brazing.

Alloy choice depends on the job requirement and on the strength required in the joint, resistance to corrosion, operating temperature, and economics. Alloys commonly used are low-fuming bronze and flux-coated low-fuming bronze.

For typical joint designs see Figure 22.2.

Use a slightly oxidizing flame. Use copper and brass flux. Mix to a paste with water. Apply to both sides of the joint. Rod can be coated with paste or heated and dipped in dry flux.

Preheating is recommended for heavy sections only.

Blowpipe tip-to-metal surface angle should be 40 to 50 degrees. The distance of the inner cone from the metal surface is 3.25 to 5.00 mm. The angle from the filler rod to the metal surface is 40 to 50 degrees.

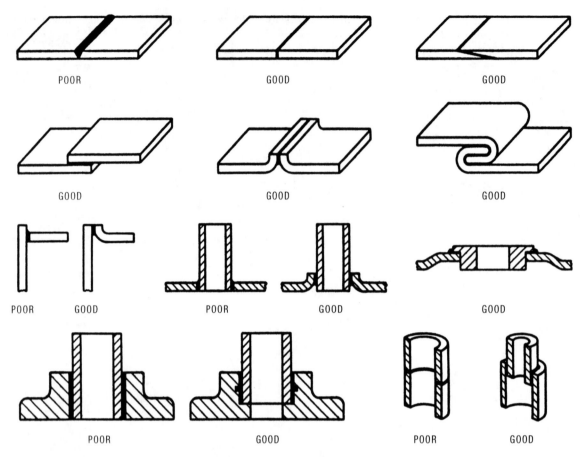

FIGURE 22.2   Good and bad joint designs for brazing.

After preheating or after the joint is raised to a temperature sufficient to permit alloying of the filler rod and copper, melt a globule of metal from the end of the rod and deposit it into the joint, wetting or tinning the surface. When tinning occurs, begin welding using the forehand technique. Do not drop filler metal on untinned surfaces.

Any of the following methods may be used to remove flux residue:

- A grinding wheel or wire brush and water
- Sandblasting
- A dilute caustic-soda dip

NICKEL AND NICKEL ALLOYS

Nickel is a versatile metal. In pure form it is used in the electroplating industry to coat metals that will be exposed to corrosive atmospheres, sea water, sulfuric acid, brine, and caustic solutions.

Nickel is added to other metals because of its ability to increase toughness and impact resistance, especially at low temperatures. It also:

- Lessens distortion in quenching
- Improves corrosion resistance
- Lowers critical temperatures
- Widens the temperature range for successful heat treatment

Nickel is used in amounts of 2 to 5 percent, often combined with other alloying elements to improve toughness. When 12 to 20 percent nickel is used in steel with low carbon content, good corrosion resistance is provided. Steel with 36 percent nickel has a thermal-expansion coefficient of almost zero.

More commonly known as invar, nickel is a metal used for measuring devices. Because of its high cost, nickel should only be used where it is uniquely effective, as in providing low-temperature impact resistance for cryogenic steels. A metric ton of nickel can sell for $30,000. It has become scarce, and deeper digging will have to be done, along with chemical mining. The demand for more power plants that use coal as a fuel is causing the shortage. These plants are required to have scrubbers to remove sulfur from their smokestack emissions. The scrubbers rely heavily on nickel to reduce corrosion and lengthen the life of the plants.

Large percentages of chromium can impart corrosion and heat resistance. But in the amounts used in low-alloy steels the effects are minor. Less than 2 percent chromium is usually required. Often chromium and nickel are used together in a ratio of about 1 part chromium to 2 parts nickel. Chromium carbides are desired for their superior wear resistance.

Nickel-based alloys are noted for their strength and corrosion resistance. This is especially true at high temperatures. Monel metal, found in an alloy form in Canada, contains about 67 percent nickel and 30 percent copper. It has been used for years in the chemical and food-processing industries because of its outstanding corrosion resistance. It probably has better overall corrosion resistance to more media than any other alloy. It is particularly resistant to salt water,and sulfuric acid, and it even resists high-velocity, high-temperature steam. For the latter reason, Monel has been used for steam-turbine blades. It can be polished for stainless-steel sinks and artwork. It shines easily and has an excellent appearance. In fact, it is often used for ornamental trim. In its common form Monel has a tensile strength from 480 to 11/0 MPa 70 to 1 1/0 ksi, depending on the amount of cold working. The elongation in 2 inches varies from 50 down to 2 percent.

There are three special grades of Monel that contain small amounts of added alloying elements:

- K-Monel contains about 3 percent aluminum and can be precipitation-hardened to a tensile strength of 1100 to 1240 MPa (160 to 180 ksi).
- H-Monel has 3 percent silicon added
- S-Monel has 4 percent silicon added.

These alloys are used for castings and can be precipitation-hardened. To improve the machining characteristics of Monel, a special free-machining alloy known as R-Monel is produced with about 0.35 percent sulfur.

Another use for nickel-based alloys is as electrical resistors. Mixed with chromium it becomes nichrome and is used as a heating element for many electrical devices.

Most of the nickel alloys are somewhat difficult to cast, but they can be forged and hot-worked. The heating, however, usually must be done in controlled atmospheres to avoid intercrystalline embrittlement. Welding of nickel or nickel alloys can be performed with little difficulty.

MAGNESIUM-BASED ALLOYS

Magnesium-based alloys are very light—about four times as light as steel or iron. That makes them ideal for aviation application.

The designation system for magnesium alloys is not as well standardized as in the case of steels or aluminums, but most producers follow a system using one or two prefix letters, two or three numerals, and a suffix letter. The prefix letters designate the two principal alloying metals according to the following format:

A	Aluminum	M	Manganese
B	Bismuth	N	Nickel
C	Copper	P	Lead
D	Cadmium	Q	Silver
E	Rare earth	R	Chromium
F	Iron	S	Silicon
H	Thorium	T	Tin
K	Zircinium	Z	Zinc
L	Beryllium		

Magnesium improves corrosion resistance, and tin improves castability. Aluminum, zinc, zirconium, and thorium promote precipitation hardening. Magnesium alloys are very machinable and are especially suitable for casting.

Spot Welding

Magnesium alloys can be spot-welded nearly as easily as aluminum. Scratch brushing or chemical cleaning is necessary before welding. Fusion welding is carried out most easily by using an inert shielding atmosphere of argon or helium gas.

Fire Hazards

Considerable misinformation has been released on magnesium and its fire-hazard characteristics. Magnesium alloys are highly combustible when they are in a finely divided form, such as a powder or fine chips, and this hazard should not be ignored.

Above 800 degrees F a noncombustible atmosphere is required to suppress burning. Castings require additional precautions due to the reactivity of magnesium with sand and water. In sheet, bar, extruded, or cast form, magnesium alloys present no fire hazard.

MULTIPLE-CHOICE EXAM

1.   Which of the following is an aluminum alloy?
     a. aluminum copper   **b.** aluminum manganese
     c. aluminum silicon   **d.** all of the above
     e. none of the above

2.   What does the first number in a classification system for wrought-aluminum alloys indicate?
     a. the alloy group   **b.** the impurity limit
     c. the purity of the aluminum **d.** none of the above

3.   Aluminum is difficult to weld because it:
     a. has a very low melting point   **b.** has a very high melting point
     c. has a low heat-loss factor   **d.** retains too much heat

4.   When aluminum is alloyed with zinc, it:
     a. produces an alloy with a low melting point
     b. produces an alloy with a high melting point
     c. produces an alloy with a shinny surface
     d. none of the above

5.   Copper is used in its pure state only in the _____ industry.
     a. electrical   **b.** jewelry
     c. construction   **d.** transportation

6.   Copper and most copper alloys can be joined by welding and _____.
     a. soldering   **b.** brazing
     c. epoxies   **d.** none of the above

7.   Welding of thick copper sections requires a high _____.
     a. heat   **b.** postheat
     c. preheat   **d.** none of the above

8. Most copper alloys even in thick sections do not require preheating because the _____ diffusivity is much lower than for copper.

    a. liquid

    b. thermal

    c. high

    d. low

9. In selecting the proper preheat for alloy welding, which factor(s) needs to be considered?

    a. the alloy being welded

    b. the base-metal thickness

    c. the overall mass of the weld

    d. all of the above

    e. none of the above

10. Aluminum-bronze and copper-_____ alloys should not be preheated.

    a. silicon

    b. nickel

    c. magnesium

    d. none of the above

11. Which is the most often used alloy material for copper?

    a. magnesium

    b. aluminum

    c. zinc

    d. none of the above

12. Deoxidized copper is a versatile _____-percent-pure copper alloy.

    a. 100

    b. 98

    c. 80

    d. 40

13. Which copper electrodes are recommended for GMAW welding of copper?

    a. Eru

    b. ERCu

    c. ECu

    d. none of the above

14. Copper sections up to 16 mm thick can be successfully welded using the _____ welding process.

    a. GTAW

    b. GMAW

    c. GMA

    d. none of the above

15. Silicon bronze is a copper-based wire recommended for _____ of copper-silicon alloys.

    a. GMAW

    b. GTAW

    c. GMA

    d. none of the above

16. Recommended shielding gases for GMA welding of copper and copper alloys are:

    a. argon-helium   **b.** argon

    c. helium   **d.** all of the above

    e. none of the above

17. In copper-silicon alloys flux residue must be removed. Which of the following is the best method to use?

    a. a dilute solution of hot caustic-soda dip

    b. a wire brushing and rinsing with hot water

    c. a wire brushing and steam

    d. all of the above

    e. none of the above

18. The braze-welding process gets its strength from the _____ strength of the filler metal deposited and the bond strength developed between the filler metal and the parent metal.

    a. tensile   **b.** welding

    c. grain   **d.** brass

19. In braze-welding which of the following is (are) used for removing flux residue?

    a. a grinding wheel   **b.** a wire brush and water

    c. sandblasting   **d.** all of the above

    e. none of the above

20. Standard solvent or _____ degreasing procedures are useful in the cleaning of copper-based metals.

    a. weak acid   **b.** alkaline

    c. hot-water   **d.** none of the above

21. The principle of brazing is to join two metals by _____ with a third metal.

    a. soldering   **b.** fusing

    c. melting   **d.** none of the above

22. To achieve an adequate bond during brazing, the following procedure(s) should be considered:

    a. joint surfaces are clean and free of oxides

    b. provision for the correct joint gap for the particular brazing filler

    c. Both a and b

    d. none of the above

23.   Nickel is added to other metals for its ability to increase _____ and impact resistance, especially at low temperatures.

    a. strength   **b.** appearance

    c. toughness   **d.** brittleness

24.   Nickel-based alloys are noted for their corrosion resistance and _____.

    a. color   **b.** weight

    c. strength   **d.** brittleness

25.   Magnesium-based alloys are very _____.

    a. heavy   **b.** light

    c. strong   **d.** weak

TRUE-FALSE EXAM

1.   Aluminum is a lightweight metal with one-third the weight of steel.

    True   False

2.   A classification system for wrought-aluminum alloys was established by the Aluminum Association in October 1954.

    True   False

3.   Pure aluminum is seldom used for casting.

    True   False

4.   Aluminum cannot be welded with oxyacetylene.

    True   False

5.   Aluminum alloyed with zinc produces an aluminum alloy with a high melting point.

    True   False

6.   Aluminum alloys along with their various tempers make up a wide and versatile range of manufacturing materials.

    True   False

7.   Copper is rarely used in its pure state except in the electrical industry.

    True   False

8. Copper and most copper alloys can be joined by welding and brazing.

True   False

9. The oxide film formed during welding should be removed with a wire brush after each weld run is deposited in copper welding.

True   False

10. Welding of thick copper sections requires a low heat.

True   False

11. Most copper alloys, even in thick sections, do not require preheating because the thermal diffusivity is much higher than for copper.

True   False

12. Aluminum-bronze and copper-nickel alloys should not be preheated.

True   False

13. ERCu copper electrodes are recommended for GMAW of copper.

True   False

14. Helium-argon mixtures are used for welding thicker sections of copper and copper alloys.

True   False

15. Braze welding and fusion welding are similar except with a filler metal that uses a lower melting point than the parent metal.

True   False

16. Alloy choice depends on the job requirements.

True   False

17. One of the oldest torches used in brazing and soldering is the blowpipe.

True   False

18. Some furnaces provide vacuums rather than shielding gases for the protection of the metal surface from atmospheric contamination.

True   False

19. A high-frequency AC generator provides the necessary heat in induction brazing.

True   False

20.   Braze welding and most brazing requires the use of a flux to prevent oxidation during the application of heat.

   True   False

21.   Nickel is used in amounts of 2 to 5 percent in combination with other alloying elements to improve toughness.

   True   False

22.   Large percentages of chromium used as an alloying agent can import corrosion resistance and heat resistance.

   True   False

23.   Monel metal contains about 67 percent nickel and 30 percent copper.

   True   False

24.   Most of the nickel alloys are somewhat difficult to cast, but they can be forged and hot-worked.

   True   False

25.   Magnesium-based alloys are very heavy in weight.

   True   False

MULTIPLE-CHOICE ANSWERS

1.	D	6.	B	11.	C	16.	D	21.	B
2.	A	7.	C	12.	B	17.	D	22.	C
3.	A	8.	B	13.	B	18.	A	23.	C
4.	A	9.	D	14.	A	19.	D	24.	C
5.	A	10.	B	15.	A	20.	B	25.	B

TRUE-FALSE ANSWERS

1.	T	8.	T	14.	T	20.	T
2.	T	9.	T	15.	T	21.	T
3.	T	10.	F	16.	T	22.	T
4.	F	11.	F	17.	T	23.	T
5.	F	12.	T	18.	T	24.	T
6.	T	13.	T	19.	T	25.	F
7.	T						

—NOTES—

Part H

JOINTS AND WELDS

Welding Licensing Exam Study Guide

—NOTES—

Chapter 23
TYPES OF JOINTS

ound weld joints are a thorough mix of the base metal and a filler rod made from a similar material. No matter the thickness of the metal, a good welded joint will be one that is fused throughout. Joining thin pieces of metal less than 1/4 inch thick makes a thoroughly welded joint less difficult. But when joining thicker metals, special joint preparation is required. Joint preparation is a necessary part of fusion welding, as it is with soldering and brazing.

Five basic types of joints in welding, shown in Figures 23.1 through 23.5, are:

- Butt joint
- Corner joint
- Edge joint
- Lap joint
- T joint

These basic joints may be used individually or in combination to produce a number of different variations.

The characteristics of a good steel-plate weld are:

- A smooth bead
- An equal junction between the two pieces of steel plate
- A bead of uniform width about 25 percent wider than the metal is thick

For instance, on steel plate 1/4 inch thick, the width of the weld bead should be about 5/16 inch. The underside of the weld should be uniform with no excess metal formations. The bead should also be covered with a thin oxide film on both sides of the joint.

METAL WARPING

When heat is applied to metal, it has a tendency to warp. This problem is more prevalent with light metals. The thinner the metal, the greater the possibility of some type of warping. Knowledge of metal and its properties aids the welder in all aspects of the joMetal, as we all know, will expand when heated and contract when cooled. This knowledge can be used to advantage when working with relatively thin sheet metal. Once the metal is welded, straightening can be done with a torch. Undesirable warpage can usually be corrected by hammering and/or heating the opposite side of the warp.

Time-proven designs for welding metal thicker than 1/4 inch are shown in Figure 23.6. Note how the joints can be prepared by a bench grinder or a hand-held grinder.

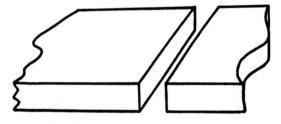

FIGURE 23.1 Butt joint.

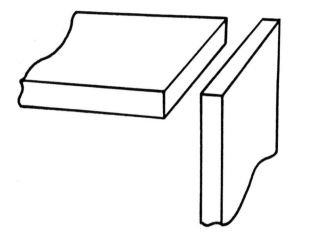

FIGURE 23.2   Corner joint.

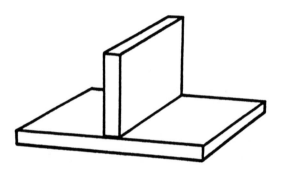

FIGURE 23.3   Edge joint.

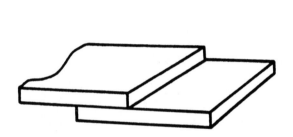

FIGURE 23.4   Lap joint.

FIGURE 23.5   T-joint.

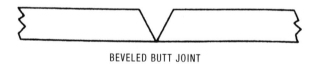

BEVELED BUTT JOINT

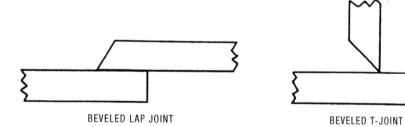

BEVELED LAP JOINT   BEVELED T-JOINT

FIGURE 23.6   Designs for welding metal thicker than 1/4 inch.

<u>**MATCHING EXAM**</u>

Match the joints with their names.

1.   _____ Butt joint

2.   _____ Corner joint

3.   _____ Edge joint

4.   _____ Lap joint

5.   _____ T joint

6.   _____ Beveled butt joint

7.   _____ Beveled lap joint

8.   _____ Beveled T joint

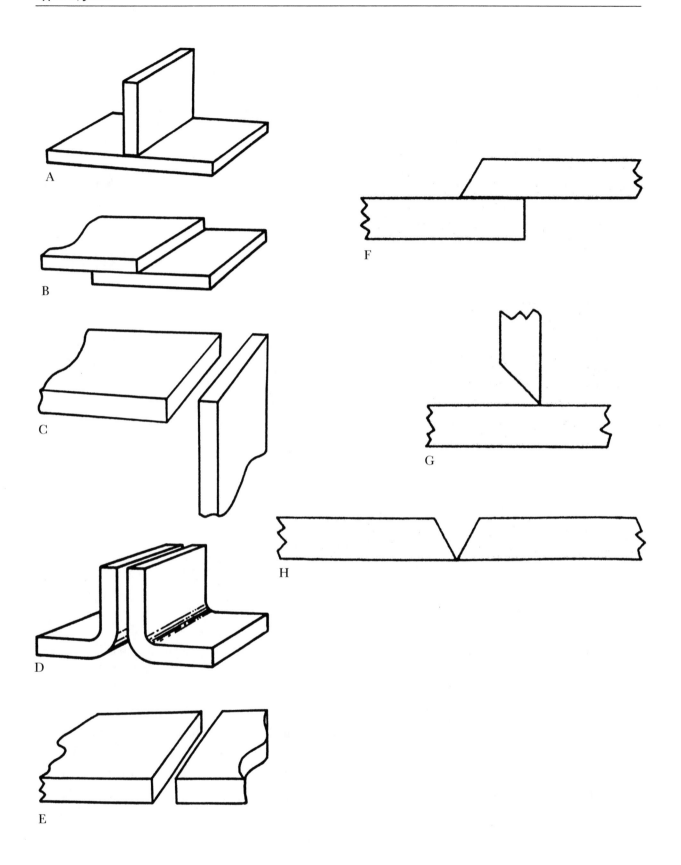

TRUE-FALSE EXAM

1. Another type of welded joint that does not necessarily require the addition of a filler rod is called the outside-corner weld.

    True   False

2. When making an outside-corner joint weld, less torch movement is necessary than usual, particularly on butt joints.

    True   False

3. Once the outside-corner weld has been completed and the metal has cooled and solidified, check the weld first for appearance and then strength.

    True   False

4. Test the strength of an outside-corner weld by trying to bend the pieces in the same way as opening a book.

    True   False

5. Another type of weld that can successfully be made without the aid of a filler rod is called a flange-joint weld.

    True   False

6. The butt joint is the easiest of all joints to make.

    True   False

7. In oxyacetylene welding, the filler rod is applied as the weld metal becomes molten and starts to flow in the butt joint.

    True   False

8. When welding the T joint, it is important not to allow too much heat to build up on the vertical piece, or a burn through may result.

    True   False

9. When welding with thick metal, some type of edge treatment must be initiated before welding begins.

    True   False

10. There are three ways that an edge of a piece of metal can be beveled.

    True   False

11. The oxyacetylene torch cannot be used to bevel and edge.

True   False

12. Use a guide on the grinder to aid in grinding a true bevel on steel plate.

True   False

13. When building up a weld bead with multilayer welding, the first layer should provide good penetration at the bottom of the V groove.

True   False

14. Most welding on steel is done manually with shielded-metal-arc (stick) electrodes.

True   False

15. Electrodes compounded to melt rapidly are called fast-fill electrodes.

True   False

16. Fast-freeze electrodes are compounded to deposit weld metal that solidifies rapidly after being melted by the arc.

True   False

17. Fast-freeze electrodes are intended specifically for welding in vertical and overhead positions.

True   False

18. A corner joint is a joint formed between two members that lie at approximately right angles to each other.

True   False

19. The butt joint is a joint formed between two members that lie approximately on the same plane.

True   False

20. The T joint is formed between two members that are located at approximately right angles to each other with the resulting connection forming a T.

True   False

MATCHING ANSWERS

1.	E	3.	D	5.	A	7.	F	8.	G
2.	C	4.	B	6.	H				

TRUE-FALSE ANSWERS

1.	T	6.	T	11.	F	16.	T
2.	T	7.	T	12.	T	17.	T
3.	T	8.	T	13.	T	18.	T
4.	T	9.	T	14.	T	19.	T
5.	T	10.	T	15.	T	20.	T

Chapter 24

WELDING POSITIONS

Welding position is the position from which the welding is performed and the filler metal is deposited in the joint. There are four basic manual welding positions:

1. Flat
2. Horizontal
3. Vertical
4. Overhead

FLAT POSITION

The flat-welding position (Figure 24.1) is one in which the welding is performed from the upper side of the joint and the face of the weld is approximately horizontal. This is the most commonly used welding position, primarily because the molten weld metal is not affected by gravity, as is the case with the other positions. In this position it is easier to deposit a uniform bead with proper penetration at a fast rate. Point the tip of the torch or the electrode downward when welding in the flat position.

HORIZONTAL POSITION

The horizontal-welding position (Figure 24.2) has two basic forms depending upon whether it is used for a fillet weld or a groove weld:

- In a fillet weld welding is performed on the upper side of an almost horizontal surface and against an almost vertical surface.
- In a groove weld the axis of the weld for the horizontal plane and the face of the weld lie in a nearly vertical plane.

Gravity in both cases can cause the molten weld metal to sag before a uniform bead with sufficient penetration can be deposited. Remember to point the tip of the torch or the electrode backward toward the weld pool when welding in the horizontal position.

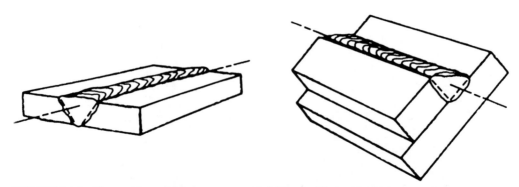

FIGURE 24.1   Flat position weld for a groove weld *(left)* and a fillet weld *(right)*.

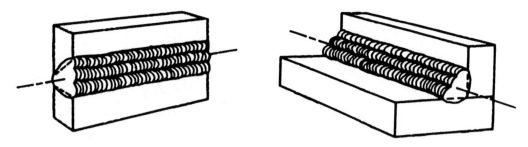

FIGURE 24.2 Flat position weld for a groove weld *(left)* and a fillet weld *(right).*

VERTICAL POSITION

The vertical-welding position (Figure 24.3) is one in which the weld material is applied to a vertical surface or one inclined 45 degrees or less to the vertical. On plates the direction of the welding can be either from the bottom to the top or from the top to the bottom. If you are welding in an upward direction in the joint, point the tip of the torch or electrode at an angle ahead of the molten weld pool. If, on the other hand, the direction of travel is downward in the joint, point the tip of the torch or electrode up and at an angle to the weld pool.

OVERHEAD POSITION

The overhead-welding position (Figure 24.4) is one in which welding is performed from the underside of the joint. It requires considerable skill and experience to lay a uniform bead with sufficient penetration. The weld metal has a tendency to sag and pull away from the joint. Point the tip of the torch or electrode upward in the joint when welding in the overhead position.

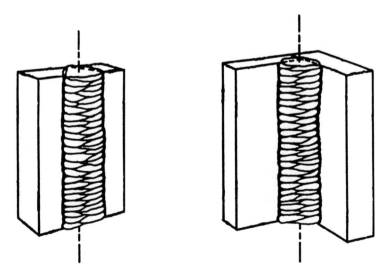

FIGURE 24.3 Vertical position weld for a groove weld *(left)* and a fillet weld *(right).*

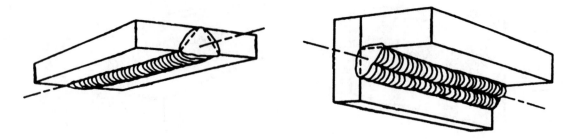

FIGURE 24.4   Overhead position for a groove weld *(left)* and a fillet weld *(right)*.

FIXED-POSITION WELDING

Sometimes the welder must be able to weld pipe that remains stationary and cannot be rolled during welding. Either vertical or horizontal fixed-position welding must be used, depending on whether the pipe is in the vertical or horizontal position. In either case the angle of the torch and filler rod (oxyacetylene welding) or the electrode (arc welding) must be constantly changed to adjust to the curvature of the pipe. Figure 24.5 illustrates the changing angle of the torch and filler rod when gas-welding a pipe in a fixed position.

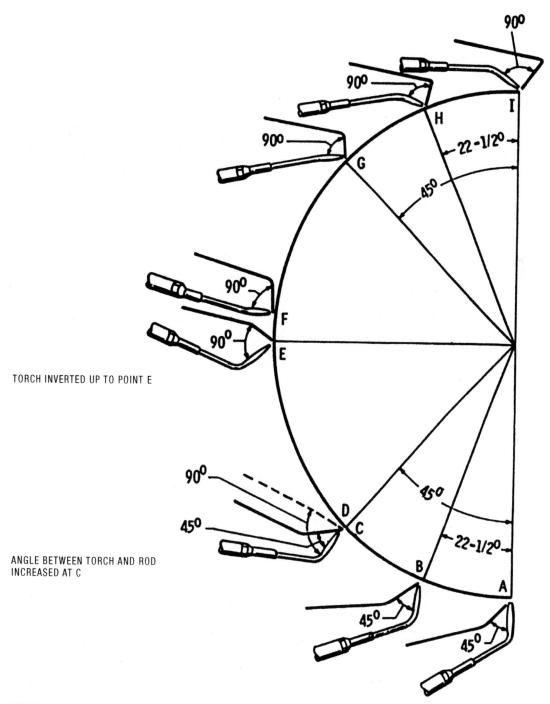

TORCH INVERTED UP TO POINT E

ANGLE BETWEEN TORCH AND ROD
INCREASED AT C

FIGURE 24.5 Various torch tip and filler rod angles for oxyacetylene welding.

MATCHING EXAM

Match the joints with their positions:

1. _____ Flat-position weld, groove

2. _____ Flat-position weld, fillet

3. _____ Horizontal-position weld, groove

4. _____ Horizontal-position weld, fillet

5. _____ Vertical-position weld, groove

6. _____ Vertical-position weld, fillet

7. _____ Overhead-position weld, groove

8. _____ Overhead-position weld, fillet

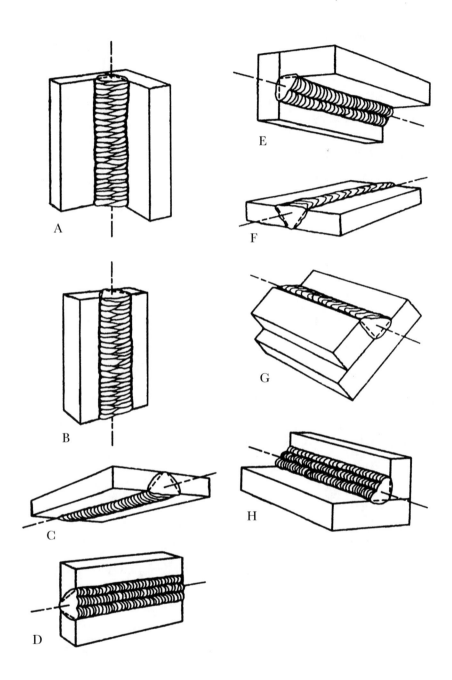

TRUE-FALSE EXAM

1.   The four positions used in manual welding are the flat position, the vertical position, the horizontal position, and the pipe-welding position.

      True   False

2.   The flat position is one in which the welding is performed from the underside of the joint and the face of the weld is approximately horizontal.

      True   False

3.   The horizontal position has two basic forms depending upon whether it is used for a fillet weld or a groove weld.

      True   False

4.   The vertical position is one in which the welding material is applied to a vertical surface or one inclined 45 degrees or less to the vertical.

      True   False

5.   Electrodes used for vertical and overhead work must provide a deposit that will stay in place and not fall out of the joint while in the molten state.

      True   False

6.   The flat position is one in which the welding is performed from the underside of the joint and the face of the weld is approximately horizontal.

      True   False

7.   The flat position is sometimes called the downward position.

      True   False

8.   The best fusion is obtained when welding in a downward horizontal-plane position on a flat piece of steel plate.

      True   False

9.   In the fillet weld the horizontal position is one in which welding is performed on the upper side of an approximately horizontal surface and against an approximately vertical surface.

      True   False

10.   In the vertical-weld position the electrode is held horizontalyl or the welding end is inclined slightly upward.

      True   False

MATCHING ANSWERS

1.	F	3.	D	5.	B	7.	C	8.	E
2.	G	4.	H	6.	A				

TRUE-FALSE ANSWERS

1.	F	4.	T	7.	T	9.	T
2.	T	5.	T	8.	T	10.	F
3.	T	6.	F				

—NOTES—

Chapter 25
TYPES OF WELDS AND WELDING

The two most common types of welds—the groove weld and the fillet weld—have a number of variations, based on differences in joint design. Other weld types include:

- Flange weld
- Surfacing weld
- Plug weld
- Slot weld
- Tack weld

GROOVE WELD

The groove weld is a weld made in a groove between two adjacent surfaces in the horizontal plane. Groove welds require edge preparation of the joining surfaces. Note the V shape at the edges where welding is to be done. Various types of groove welds are shown in Figure 25.1.

FILLET WELD

The fillet weld is a weld joining two surfaces positioned at right angles to each other in a lap, T-, or corner joint. The cross section of a fillet weld is approximately triangular with either a convex or concave face. See Figures 25.2 and 25.3.

A full fillet weld is one whose size is equal to the thickness of the thinner of the two members being joined.

Staggered intermittent fillet welds (Figure 25.4) consist of two lines of intermittent fillet welds on a joint. The weld increments in one line are staggered with respect to those in the other. Chain intermittent welds (Figure 25.5), on the other hand, consist of two lines of intermittent fillet welds in which the weld increments are located opposite one another.

FLANGE WELD

A flange weld is a weld made on the edges of two or more members to be joined, at least one of which has been flanged or turned back at right angles to the surface for reinforcement. The corner flange weld (Figure 25.6) is an example of this type of weld in which the edge of only one member has been flanged. In the case of the edge flange weld (Figure 25.7) the edges of both members have been flanged. The flange weld is frequently used in sheet-metal work. Filler metals are not necessary.

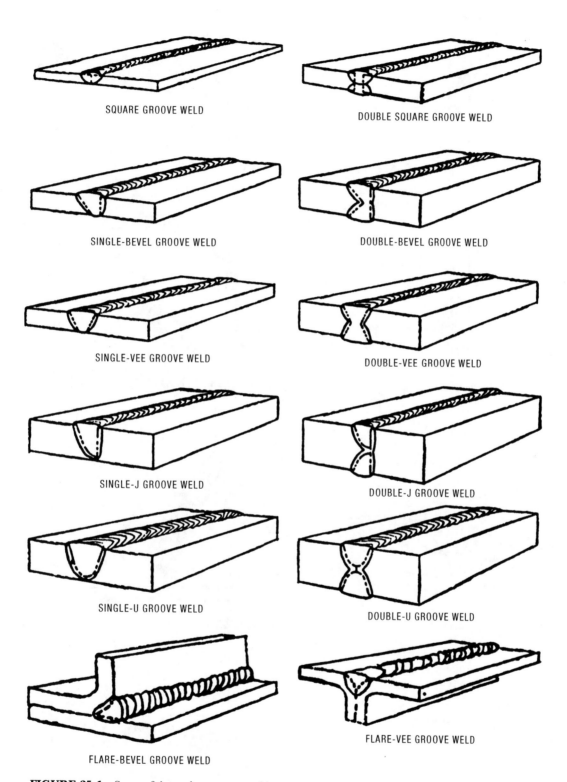

SQUARE GROOVE WELD

DOUBLE SQUARE GROOVE WELD

SINGLE-BEVEL GROOVE WELD

DOUBLE-BEVEL GROOVE WELD

SINGLE-VEE GROOVE WELD

DOUBLE-VEE GROOVE WELD

SINGLE-J GROOVE WELD

DOUBLE-J GROOVE WELD

SINGLE-U GROOVE WELD

DOUBLE-U GROOVE WELD

FLARE-BEVEL GROOVE WELD

FLARE-VEE GROOVE WELD

FIGURE 25.1   Some of the various groove weld types.

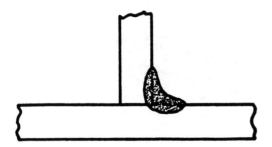

FIGURE 25.2   Weld using a concave fillet.

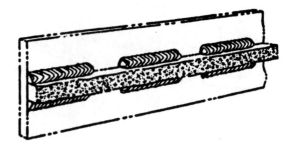

FIGURE 25.5   A chain intermittent weld.

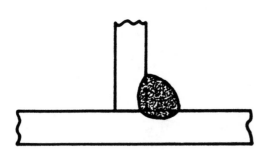

FIGURE 25.3   Weld using a convex fillet.

FIGURE 25.6   A corner flange weld.

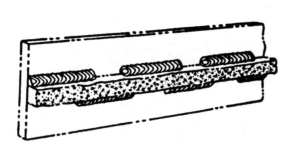

FIGURE 25.4   A staggered intermittent fillet weld.

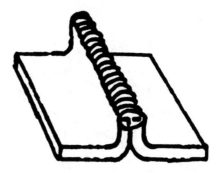

FIGURE 25.7   An edge flange weld.

FIGURE 25.8   A surfacing weld.

FIGURE 25.9   A plug weld.

FIGURE 25.10   A slot weld.

SURFACING WELD

A surfacing weld is a type of weld deposited on an unbroken surface to obtain certain properties or dimensions (Figure 25.8). These welds consist of one or more string or weave beads. A surface weld is sometimes referred to as a bead weld.

PLUG WELD

A plug weld (Figure 25.9) is a circular weld made through a hole in one member of a lap or T-joint joining that member to the other.

SLOT WELD

A slot weld (Figure 25.10) is used with the same type of joints (lap or T-) but is distinguished by its more elongated hole.

TACK WELD

A tack weld is a temporary weld used to hold two parts in position until a more permanent weld can be made (Figure 25.11). It is not intended for use as a permanent weld.

SPOT WELDS

There are other types of welds to be considered when looking at the entire welding industry. One of the most often used in the automobile industry is the spot weld. It is used to hold together the car or truck body.

Resistance spot welding is one of the oldest of the electric welding processes in use today. The weld is made by a combination of heat, pressure, and time. As the name implies, it is the resistance of the material to be welded to the current flow that causes a localized heating in the part. The pressure exerted by the tongs and electrode tips through which the current flows holds the parts to be welded in intimate contact before,

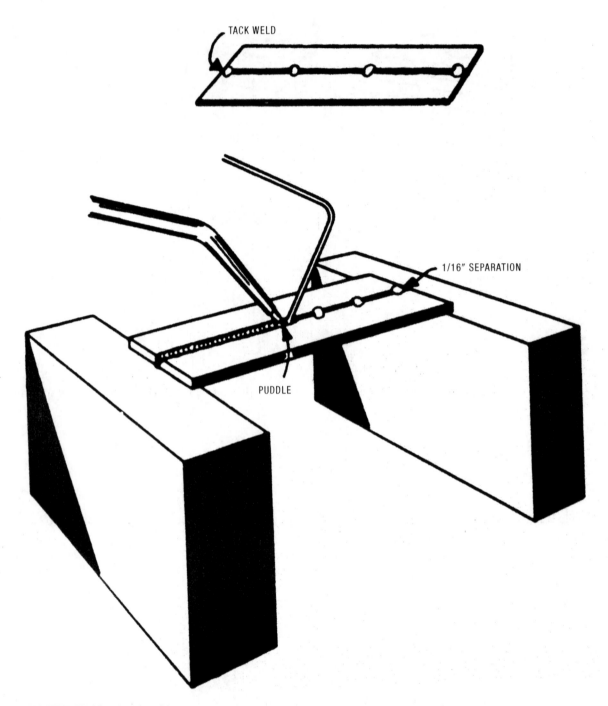

FIGURE 25.11 A tack weld.

during, and after the welding current cycle. The required amount of time current flows in the joint is determined by material thickness and type, the amount of current flowing, and the cross-sectional areas of the welding tip contact surfaces. (See Figure 25.12.)

The gas-tungsten-arc spot is made from one side only. The resistance weld is normally made with electrodes on each side of the workpiece. Resistance spot welds may be made with the workpiece in any position.

The resistance-spot-weld nugget is formed when the interface of the weld joint is heated due to the resistance of the joint surfaces to electric current. In all cases, of course, the current must flow or the weld cannot

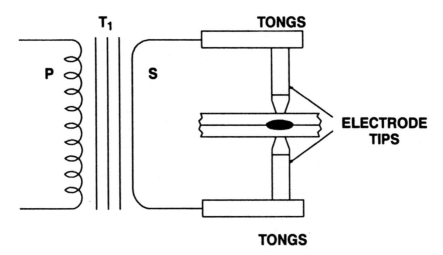

FIGURE 25.12 Resistance spot welding. P is for primary and S is for secondary of the transformer, T_1. *(Courtesy of Miller Welds)*

be made. The pressure of the electrode tips on the workpieces holds the part in close and intimate contact during the making of the weld. Remember, however, that resistance-spot-welding machines are not designed as force clamps to pull the workpieces together for welding. (See Figure 25.13.)

As with any type of equipment, the simplest of welding machines, the electric spot welder, may require troubleshooting from time to time. Table 25.1 lists some of the possibilities for trouble and their remedies.

SEAM WELDING

Seam welding is similar to spot welding. Note the circular electrode used to produce the effect of a continuous seam weld. (See Figure 25.14.)

FLASH WELDING

Flash welding is a resistance-welding process in which the heat is applied before the two parts are pressed together. When the pressure is applied, molten metal is squeezed out of the weld. (See Figure 25.15.)

In upset welding the heat is applied after the two metal pieces are clamped in the electrodes. (See Figure 25.16.)

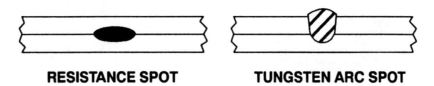

RESISTANCE SPOT   **TUNGSTEN ARC SPOT**

FIGURE 25.13 Comparison of resistance and TIG spot welds. *(Courtesy of Miller Welds)*

TABLE 25.1   Troubleshooting spot welders. *(Courtesy of Miller Welds)*

Trouble	Remedy
Tips overheating.	Not enough tong pressure. Increase tong pressure.
	Weld time too long. Reduce weld time.
	Material too thick for the spot welding machine.
Tips arcing on material.	Not enough tong pressure. Increase tong pressure.
	Tips not aligned correctly. Realign tips or dress tips to proper diameter.
	Base material may be welded to tips causing high resistance and poor electrical current flow. Clean or dress tips.
Spatter or molten material being expelled out during welding operation.	Incorrect tip alignment. Dress tips so that they align and are flat on the material.
	Excessive tong pressure. Reduce tong pressure.
	Output amperage too high. Reduce amperage setting, if applicable (not available on air-cooled models).
	Weld time too long. Reduce weld time.
Inconsistent weld nugget.	Inconsistent weld time. Install a weld timer, if applicable.
	Not enough tong pressure. Increase tong pressure.
Hole in middle of weld.	Contact area of tips is too large. Change to a smaller tip diameter or dress tips back to original diameter.
Poor weld or no weld at tips.	Material too thick for spot welding machine. Check that material thickness is within capacity of spot welding machine.
	Tongs are too long. Reduce tong length.
	Remove coating from material for intimate contact between pieces. Remove oxides and chemical compounds including galvanized coating.

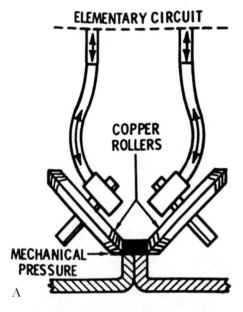
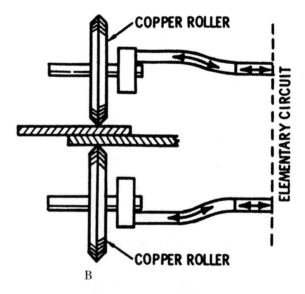

FIGURE 25.14   Seam welding.

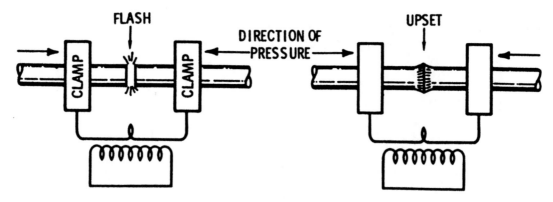

FIGURE 25.15   Flash welding.

In butt welding the parts are placed end to end, heated electrically, and then pressed together with enough pressure to form a weld.

Another type of welding is percussion welding, also called percussive welding. It is a resistance-welding process whereby the electrical energy is stored up and released with a sudden explosive impact. Fusion occurs simultaneously over the entire contact area.

Induction welding is an electrical welding process that uses the resistance of metal to the flow of an induced electrical current to produce the heat. It is in the application of the electrical current that induction welding differs from resistance welding. In resistance welding the work is placed between two electrodes. The electric current passes through the metal on its path from one electrode to the other. In induction welding, however, the work is placed in a radio-frequency field. (Radio frequency is anything above the audible range of 16,000 Hz). The rf field is produced within coils that are designed to approximate the shape of the work. It induces an electric current in the pieces to be welded, and they heat up. This type of induction welding is essentially a production welding process.

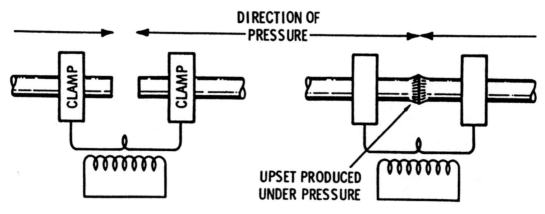

FIGURE 25.16   Upset welding.

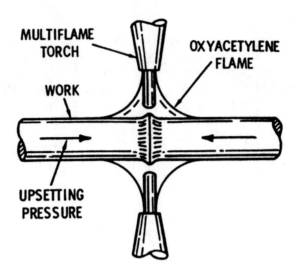

FIGURE 25.17   Pressure and gas used for welding.
The bulging ring around the joint is later ground down.

GAS WELDING

In some instances it is less expensive to use gas as a source of heat. The gas usually consists of an acetylene and oxygen mixture. Gas welding is a puddling process in which the molten metal that forms the weld is in a small pool over which the flame constantly plays. More than one torch may be used to produce the heat. (See Figure 25.17.)

FORGE OR FURNACE HEATING

Forge welding is the same process as that utilized by the village blacksmith. The metal is heated to the proper temperature and then hammered or pressured together with the hammer on the anvil. There are three forms of forge welding:

- Hammer welding
- Die welding
- Roll welding

Today the various forge-welding processes are done primarily with machines and are essentially die-forming processes.

ELECTRON BEAMS

A stream of concentrated electrons is produced by an electron gun and accelerated to gain momentum so that the heat produced when the electrons hit the workpiece produces high heat. This type of welding is usually done in a vacuum and produces very little if any contamination.

CHEMICAL REACTIONS

Thermal welding refers to a group of processes whereby the high temperatures necessary for fusion are produced by means of a chemical reaction between a metal oxide and aluminum. The chemical reaction produces a quantity of superheated molten steel. Not only does this heat the base metals, but it also acts as the filler metal in the mold. This type of welding is used mostly in the joining of rails and reinforcing bars.

SOUND

Ultrasonic welding is a process that uses high-intensity vibrations slightly above human hearing range to produce the temperatures necessary for fusion. This process is used primarily in welding aluminum, magnesium, and their alloys but is easily adapted to welding plastics.

   This type of welding is limited to thin metals—foil and wire. The pieces to be welded are placed between a tip and a reflecting anvil, thereby concentrating the vibratory energy within the work. Either stationary tips for spot welds or rotating disks,for seam welds can be used.

OTHER PROCESSES

There are more welding processes (some not yet discovered). To list a few:

- Light welding
- Flow welding
- Cold welding
- Friction welding
- Plastics welding

MATCHING EXAM

Match the type of weld with its name:

1.   _____ Flare-bevel groove

2.   _____ Square groove

3.   _____ Single-U groove

4.   _____ Double-bevel groove

5.   _____ Single-J groove

6.   _____ Double-V groove

7.   _____ Single-V groove

8.   _____ Double-J groove

9.   _____ Single-bevel groove

10. _____ Double-U groove

11. _____ Double-square groove

12. _____ Flare-V groove

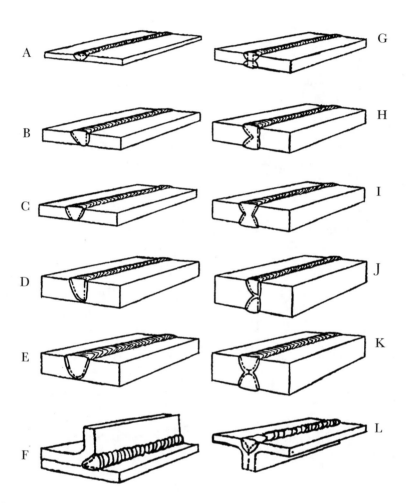

TRUE-FALSE EXAM

1.   Two of the most common types of welds are groove welds and fillet welds.

           True   False

2.   Groove welds require edge preparation of the joining surfaces.

           True   False

3.   The cross section of a fillet weld is approximately triangular with either a convex or concave face.

           True   False

4.   A flange weld is a weld made on the edges of two or more members to be joined, at least one of which has been flanged or turned back at right angles to the surface.

           True   False

5.   The flange weld is frequently used in heavy steel work.

           True   False

6.   A surfacing weld is a type of weld deposited on a broken surface to obtain certain properties.

           True   False

7.   A plug weld is a circular weld made through a hole in one member of a lap or T-joint, joining that member to the other.

           True   False

8.   A slot weld is used with the same type of joints (lap or T-) but is distinguished from the plug weld by it longer elongated hole.

           True   False

9.   A tack weld is a permanent weld used to hold two parts in position until a more permanent weld can be made.

           True   False

10.   Resistance welding is one of the oldest of the electric welding processes in use today.

           True   False

11.   The gas-tungsten-arc spot weld is made from one side only.

           True   False

12. The staggered intermitten fillet welds consist of two lines of intermitten welds on a joint.

True   False

13. The resistance-spot-weld nugget is formed when the interface of the weld joint is heated due to the resistance of the joint surfaces to electric current.

True   False

14. Resistance-spot-welding machines are designed as force clamps to pull the workpieces together for welding.

True   False

15. In upset welding the heat is applied after the two metal pieces are clamped in the electrodes.

True   False

16. In butt welding the parts are placed end to end, heated electrically, and then pressed together with enough pressure to form a weld.

True   False

17. Induction welding is an electrical welding process that uses the resistance of metal to the flow of an induced electrical current to produce the heat.

True   False

18. Forge welding is the same in principle as the process used by the village blacksmith years ago.

True   False

19. Today the various forms of forge welding are done primarily with machines and are essentially die-forming processes.

True   False

20. Ultrasonic welding is a process that uses high-intensity vibrations to produce the temperatures necessary for fusion.

True   False

MATCHING ANSWERS

1. F	4. H	7. C	9. B	11. G
2. A	5. D	8. J	10. K	12. L
3. E	6. I			

TRUE-FALSE ANSWERS

1. T	6. F	11. T	16. T
2. T	7. T	12. T	17. T
3. T	8. T	13. T	18. T
4. T	9. F	14. F	19. T
5. F	10. T	15. T	20. T

Chapter 26

WELDING TIPS AND TESTS

Inspection and testing are part of the welding trade. Many lives and buildings depend on a good weld performing the way it was designed to perform. In order to have a good weld that can be depended upon to serve its intended purpose, it must be inspected carefully, thoroughly, and frequently. The American Welding Society (AWS) has a testing program to certify those who have certain skills in this area of specialization. A good weld has superior strength to the base (parent) metal surrounding it. When an overload force is applied to the joint, the base metal will most likely give before the weld does. A new welder can consistently produce welds of high quality if care is taken in both the preparation of the weld and the welding procedure.

WELDING TIPS

Following are some useful tips for your welding projects:

- Correctly identify the base (parent) metal of the joint to be welded.

- Take into consideration the type of base metal as well as the design and position of the joint. Select a welding process capable of meeting these requirements.

- Use welding equipment and supplies of the highest quality. To save expense at this point is to run the risk of producing poor-quality welds.

- Choose the most suitable electrode or filler metal for the weld. This will depend on the identification of the base metal and the requirements of the welding job.

- Check and meet (or slightly exceed) the size requirements for the weld. If the metal is too thick for adequate penetration, bevel the edges to be joined.

- Whenever possible, thoroughly clean the surface of all contaminants (dirt, grease, oil, etc.). If these remain on the surface, they will likely be included in the weld and thereby weaken it.

- Begin welding by laying the initial bead. Clean the bead, remove any weld slag, and begin the next bead (or pass) until the weld is completed.

- Examine the weld. If it is defective, remove it immediately and begin again. Never attempt to weld over a porous weld or a cracked weld.

- The finished weld should be clean and have a good appearance. All undercutting and overlapping should be repaired.

- Some metals require preheating to reduce the possibility of cracking and the formation of residual stresses. This requires a correct identification of the metal being welded. If preheating is required, use the preheat method specified. Note: You should never weld on metal surfaces below 32 degrees F without first preheating.

- Examine the finished weld and correct any surface defects. The finished appearance of the weld should be smooth. If undercuts or overlaps are discovered, repair them.

TESTING

Nondestructive and destructive tests are the two general categories for welding joints. Almost every type of nondestructive test requires trained, qualified personnel and special equipment for its effective use. Visual inspection is probably the only exception to this rule.

Nondestructive Testing

The nondestructive testing categories consist of the following:

- Visual inspection
- Magnetic particles
- Eddy currents
- Ultrasound
- Radiography
- Penetration

In some instances visual inspection is not included in the nondestructive category. Since there is no way to internally inspect the weld, a visual inspection should include a close examination of the external part of the joint. Some of the things to look for in the inspection are:

- Weld-bead size
- Weld-bead contour
- Crater deficiencies
- Degree of fusion (surface)
- Absence or presence of cracks under cuttings and shrinkage cavities

Destructive Testing

Destructive tests are used to determine the properties of a weld but in such a way that the weld itself is usually destroyed. A good example of destructive testing is in the assembly of automobile bodies. Using an air-driven chisel and hammer, testers try to separate the spot welds that hold the body parts together. Of course, the body is no longer usable, so it is scrapped. The destructive tests can be categorized into the following:

- Bend test
- Etch test
- Impact test
- Hardness test
- Tensile test
- Nick-break test

TRUE-FALSE EXAM

1.   Inspection and testing are part of the welding trade.
     True   False

2.   Certification of inspectors is done by the American Welding Society.
     True   False

3.   A good weld has superior strength to the base metal surrounding it.
     True   False

4.   In order to make a good weld, it is important to identify the base metal of the joint to be welded.
     True   False

5.   Choosing the most suitable electrode or filler metal for the weld is not necessary.
     True   False

6.   Some metals require preheating to reduce the possibility of cracking and the formation of residual stresses.
     True   False

7.   Never attempt to weld over a porous or a cracked weld.
     True   False

8.   You should never weld on metal surfaces below 32 degrees F.
     True   False

9.   The destructive bend test may be used to determine a number of weld properties including ductility, weld penetration, tensile strength, and fusion.
     True   False

10. Destructive tests are used to determine the properties of a weld but in such a way that the weld itself is usually destroyed.
     True   False

11. The etch test is nondestructive.

    True   False

12. The eddy-current test is destructive.

    True   False

13. The root-bend tests are used primarily to determine the degree of weld penetration.

    True   False

14. The eddy-current testing method uses electromagnetic energy for detecting and locating weld defects.

    True   False

15. Magnetic-particle testing is a form of weld inspection in which surface or near-surface flaws are located by means of an induced magnetic field.

    True   False

TRUE-FALSE ANSWERS

1.	T	5.	F	9.	T	13.	T
2.	T	6.	T	10.	T	14.	T
3.	T	7.	T	11.	F	15.	T
4.	T	8.	T	12.	F		

Chapter 27
MATHEMATICS

Welding Licensing Exam Study Guide

rithmetic is the art of dealing with real numbers. It deals with the addition, subtraction, multiplication, and division of numbers. If you want to extend the manipulation of numbers and observe the other properties that numbers possess, it is necessary to take a look at a higher step on the ladder of mathematics. Algebra is that next step.

Algebra is the branch of mathematics that deals with the relations and properties of numbers by means of letters, signs of operation, and other symbols, including solutions of equations, polynomials, continued fractions, and other mathematical phenomena. Algebra plays a fundamental role in the solution of many welding problems. It can be used to express abstract concepts in simple terms. The abstract concepts are important in working with metals and welding.

Math requirements for welders are not very difficult to master. But some welders have trouble with the math portions of their certification exam. Many test facilities will allow the use of a calculator. Using the calculator makes the task of solving problems rather easy. First learn to use the square root ($\sqrt{}$) by putting in the number 4 and then taking the square root of it. The answer is, of course, 2. Now take the square root of 25. The answer is 5. Now take a look at the calculator once more. Find the x^2 key. Put in the number 2 and hit the square key (x^2). The answer is 4. Now that the calculator is warmed up and you are ready to go, take a look at the problems that follow and check your answer with your calculator.

TABLE 27.1   Temperature conversion.

$°F = 1.8 \times °C + 32$
$°C = 0.5555555 \times °F - 32$
$°C = °F - 32 \div 1.8$
$°F = °C \times 1.8 + 32$

TABLE 27.2   Cube roots of fractions.

Fraction	Cube root
1/8	0.5000
1/4	0.6300
3/8	0.7211
1/2	0.7937
5/8	0.8550
3/4	0.9086
7/8	0.9565

TABLE 27.3   Square feet to square meters.

Square feet	Square meters
1	0.925
2	0.1850
3	0.2775
4	0.3700
5	0.4650
6	0.5550
7	0.6475
8	0.7400
9	0.8325
10	0.9250
25	2.315
50	4.65
100	9.65

TABLE 27.4 Cubes of numbers.

Number	Cube	Number	Cube	Number	Cube
1	1	36	46,656	71	357,911
2	8	37	50,653	72	373,248
3	27	38	54,872	73	389,017
4	64	39	59,319	74	405,224
5	125	40	64,000	75	421,875
6	216	41	68,921	76	438,976
7	343	42	74,088	77	456,533
8	512	43	79,507	78	474,552
9	729	44	85,184	79	493,039
10	1,000	45	91,125	80	512,000
11	1,331	46	97,336	81	531,441
12	1,728	47	103,823	82	551,368
13	2,197	48	110,592	83	571,787
14	2,477	49	117,649	84	592,704
15	3,375	50	125,000	85	614,125
16	4,096	51	132,651	86	636,056
17	4,913	52	140,608	87	658,503
18	5,832	53	148,877	88	681,472
19	6,859	54	157,464	89	704,969
20	8,000	55	166,375	90	729,000
21	9,621	56	175,616	91	753,571
22	10,648	57	185,193	92	778,688
23	12,167	58	195,112	93	804,357
24	13,824	59	205,379	94	830,584
25	15,625	60	216,000	95	857,375
26	17,576	61	226,981	96	884,736
27	19,683	62	238,328	97	912,673
28	21,952	63	250,047	98	941,192
29	24,389	64	262,144	99	970,299
30	27,000	65	274,625	100	1,000,000
31	29,791	66	287,496		
32	32,768	67	300,763		
33	35,937	68	314,432		
34	39,304	69	328,500		
35	42,875	70	343,000		

TABLE 27.5 Square roots of numbers.

Number	Square root	Number	Square root	Number	Square root
1	1.00000	36	6.00000	71	8.42614
2	1.41421	37	6.08276	72	8.48528
3	1.73205	38	6.16441	73	8.54400
4	2.00000	39	6.24499	74	8.60232
5	2.23606	40	6.32455	75	8.66025
6	2.44948	41	6.40312	76	8.71779
7	2.64575	42	6.48074	77	8.77496
8	2.82842	43	6.55743	78	8.83176
9	3.00000	44	6.63324	79	8.88819
10	3.16227	45	6.70820	80	8.94427
11	3.31662	46	6.78233	81	9.00000
12	3.46410	47	6.85565	82	9.05538
13	3.60555	48	6.92820	83	9.11043
14	3.74165	49	7.00000	84	9.16515
15	3.87298	50	7.07106	85	9.21954
16	4.00000	51	7.14142	86	9.27361
17	4.12310	52	7.21110	87	9.32737
18	4.24264	53	7.28010	88	9.38083
19	4.35889	54	7.34846	89	9.43398
20	4.47213	55	7.41619	90	9.48683
21	4.58257	56	7.48331	91	9.53939
22	4.69041	57	7.54983	92	9.59166
23	4.79583	58	7.61577	93	9.64365
24	4.89897	59	7.68114	94	9.69535
25	5.00000	60	7.74596	95	9.74679
26	5.09901	61	7.81024	96	9.79795
27	5.19615	62	7.87400	97	9.84885
28	5.29150	63	7.93725	98	9.89949
29	5.38516	64	8.00000	99	9.94987
30	5.47722	65	8.06225	100	10.00000
31	5.56776	66	8.12403		
32	5.65685	67	8.18535		
33	5.74456	68	8.24621		
34	5.83095	69	8.30662		
35	5.91607	70	8.36660		

TABLE 27.6   Circumference of circle.

Diameter	Circumference	Diameter	Circumference
⅛	0.3927	10	31.41
¼	0.7854	10½	32.98
⅜	1.178	11	34.55
½	1.570	11½	36.12
⅝	1.963	12	37.69
¾	2.356	12½	39.27
⅞	2.748	13	40.84
1	3.141	13½	42.41
1⅛	3.534	14	43.98
1¼	3.927	14½	45.55
1⅜	4.319	15	47.12
1½	4.712	15½	48.69
1⅝	5.105	16	50.26
1¾	5.497	16½	51.83
1⅞	5.890	17	53.40
2	6.283	17½	54.97
2¼	7.068	18	56.54
2½	7.854	18½	58.11
2¾	8.639	19	56.69
3	9.424	19½	61.26
3¼	10.21	20	62.83
3½	10.99	20½	64.40
3¾	11.78	21	65.97
4	12.56	21½	67.54
4½	14.13	22	69.11
5	15.70	22½	70.68
5½	17.27	23	72.25
6	18.84	23½	73.82
6½	20.42	24	75.39
7	21.99	24½	76.96
7½	23.56	25	78.54
8	25.13	26	81.68
8½	26.70	27	84.82
9	28.27	28	87.96
9½	29.84	29	91.10
		30	94.24

TABLE 27.7   Area of circle.

Diameter	Area	Diameter	Area
⅛	0.0123	10	78.54
¼	0.0491	10½	86.59
⅜	0.1104	11	95.03
½	0.1963	11½	103.86
⅝	0.3068	12	113.09
¾	0.4418	12½	122.71
⅞	0.6013	13	132.73
1	0.7854	13½	143.13
1⅛	0.9940	14	153.93
1¼	1.227	14½	165.13
1⅜	1.484	15	176.71
1½	1.767	15½	188.69
1⅝	2.073	16	201.06
1¾	2.405	16½	213.82
1⅞	2.761	17	226.98
2	3.141	17½	240.52
2¼	3.976	18	254.46
2½	4.908	18½	268.80
2¾	5.939	19	283.52
3	7.068	19½	298.60
3¼	8.295	20	314.16
3½	9.621	20½	330.06
3¾	11.044	21	346.36
4	12.566	21½	363.05
4½	15.904	22	380.13
5	19.635	22½	397.60
5½	23.758	23	415.47
6	28.274	23½	433.73
6½	33.183	24	452.39
7	38.484	24½	471.43
7½	44.178	25	490.87
8	50.265	26	530.93
8½	56.745	27	572.55
9	63.617	28	615.75
9½	70.882	29	660.52
		30	706.86

TABLE 27.8 Area and other formulas.

Parallelogram	Area = base × distance between the two parallel sides
Pyramid	Area = ½ perimeter of base × slant height + area of base
	Volume = area of base × ⅓ of the altitude
Rectangle	Area = length × width
Rectangular prism	Volume = width × height × length
Sphere	Area of surface = diameter × diameter × 3.1416
	Side of inscribed cube = radius × 1.547
	Volume = diameter × diameter × diameter × 0.5236
Square	Area = length × width
Triangle	Area = one-half of height times base
Trapezoid	Area = one-half of the sum of the parallel sides × the height
Cone	Area of surface = one-half of circumference of base × slant height + area of base
	Volume = diameter × diameter × 0.7854 × one-third of the altitude
Cube	Volume = width × height × length
Ellipse	Area = short diameter × long diameter × 0.7854
Cylinder	Area of surface = diameter × 3.1416 × length + area of the two bases
	Area of base = diameter × diameter × 0.7854
	Area of base = volume ÷ length
	Length = volume ÷ area of base
	Volume = length × area of base
	Capacity in gallons = volume in inches ÷ 231
	Capacity of gallons = diameter × diameter × length × 0.0034
	Capacity in gallons = volume in feet × 7.48
Circle	Circumference = diameter × 3.1416
	Circumference = radius × 6.2832
	Diameter = radius × 2
	Diameter = square root of = (area ÷ 0.7854)
	Diameter = square root of area × 1.1233

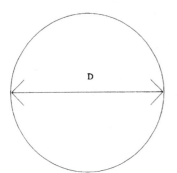

FIGURE 27.1   Diameter of a circle.

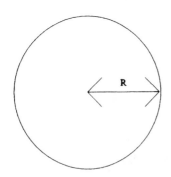

FIGURE 27.2   Radius of a circle.

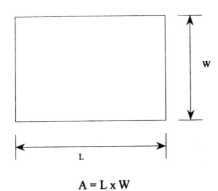

$$A = L \times W$$

FIGURE 27.3   Angle of a rectangle.

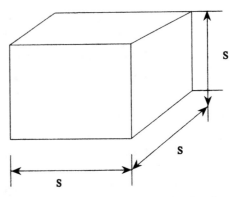

Volume is S cubed or V = S x S x S

FIGURE 27.4   Volume of a cube.

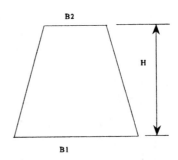

**A = Area, B1 = First Base,
   B2 = Second Base, H = Height**

**To find the area of a trapezoid,
   use the following formula:**

$$A = \frac{1}{2}(B1 + B2)\,H$$

Note: Perform math function in parenthesis first.

FIGURE 27.5   Area of a trapezoid.

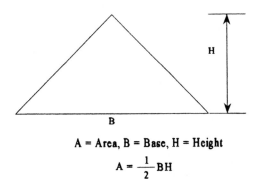

A = Area, B = Base, H = Height

$$A = \frac{1}{2}BH$$

FIGURE 27.6 Area of a triangle.

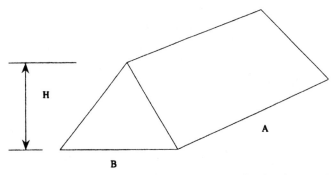

V = Volume, A = Length of Prism, B = Base of Triangle, H = Height

$$V = \frac{ABH}{2}$$

FIGURE 27.7 Area of a triangular prism.

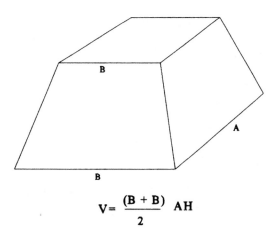

$$V = \frac{(B + B)}{2}AH$$

FIGURE 27.8 Volume of a trapezoidal prism.

MULTIPLE-CHOICE EXAM

1.   What is the square of the number 6?

     a. 36   **b.** 12

     c. 24   **d.** 30

2.   What is the square root of 49?

     a. 8   **b.** 7

     c. 9   **d.** 6

3.   How many square inches are in in 16 square feet?

     a. 1000   **b.** 2040

     c. 2304   **d.** 256

4.   How many cubic feet are in 1728 cubic inches?

     a. 10   **b.** 8

     c. 7   **d.** 1

5.   How many square feet are in a square yard?

     a. 36   **b.** 10

     c. 12   **d.** 9

6.   How many square feet are in a 20 × 20 foot steel plate?

     a. 100   **b.** 200

     c. 300   **d.** 400

7.   How many feet are in one mile?

     a. 5000   **b.** 6000

     c. 5280   **d.** 4500

8.   What is the value of the Greek letter pi in mathematical equations?

     a. 3.0   **b.** 3.14152654

     c. 2.1456890   **d.** 2.0

9. What is the approximate square root of 34?

   a. 5.831   **b.** 5.021

   c. 6.0123   **d.** 4.931

10. When reading a blueprint, you find a surface that measures 6 inches on your scale. The scale being used is: 1/4 inch = 1 foot. What is the length of that surface?

    a. 48 feet   **b.** 32 feet

    c. 24 feet   **d.** 72 feet

11. What is the volume of a piece of steel that measures 3 inches × 4 inches × 2 feet?

    a. 288 cubic inches   **b.** 24 cubic feet

    c. 24 cubic inches   **d.** 144 cubic inches

12. If you are reading a set of blueprints drawn to a scale of 1/8 inch per foot, how long is a steel rod that scales out at 4 inches?

    a. 2 feet   **b.** 4 feet

    c. 96 feet   **d.** 32 feet

TRUE-FALSE EXAM

1. The formula to find the cube of the number 4 is 4 × 4 × 4.

   True   False

2. To find the volume of a solid piece of steel, multiply length × width × thickness.

   True   False

3. An octagon has eight sides.

   True   False

4. You can find the diameter of a circle by dividing the circumference by π.

   True   False

5. The circumference of a circle can be found by multiplying the diameter by π.

   True   False

6.   To find the area of a square, multiply the length × width × thickness.

   True   False

7.   The volume of a sphere is found by multiplying the diameter cubed × 0.5236.

   True   False

8.   A pentagon has five sides.

   True   False

9.   To find the area of a triangle, multiply one-half of height × the base.

   True   False

10.   To find the volume of a pyramid, multiply the area of the base × one-third the altitude.

   True   False

11.   To find the volume of a cube, multiply length × width × height.

   True   False

12.   To find the volume of a cone, multiply π × radius squared × one-third the height.

   True   False

13.   Quadrilaterals are figures with three lines.

   True   False

14.   To convert Celsius to Fahrenheit, multiply Celsius temperature × 1.8 and add 32.

   True   False

15.   To convert Fahrenheit to Celsius, multiply 0.55555555 × Fahrenheit temperature and subtract 32.

   True   False

16.   To find the area of a triangular prism, multiply width × height × length and divide by 2.

   True   False

17.   Is the following true or false: 9 square meters is equal to 9 square feet.

   True   False

18. The cube root of 1/2 is 0.8550.

    True   False

19. The square root of 12 is 3.46410.

    True   False

20. If the diameter of a circle is 5, the circumference is 15.7.

    True   False

MULTIPLE-CHOICE ANSWERS

1. A	4. D	7. C	9. A	11. A
2. B	5. D	8. B	10. C	12. D
3. C	6. D			

TRUE-FALSE ANSWERS

1. T	6. F	11. T	16. T
2. T	7. T	12. T	17. F
3. T	8. T	13. F	18. F
4. T	9. F	14. T	19. T
5. T	10. T	15. T	20. T

Chapter 28

READING BLUEPRINTS

usion and resistance welding are the two main categories for welding processes. Fusion welding involves the melting of a metal rod, called a welding rod, and combining it with the metal parts. Once the melted rod cools, the parts are permanently joined. The rod can be melted by electricity or gas. Resistance welding involves passing an electric current under pressure through the area to be welded. The resistance to current flow causes the heat that melts the metal parts. Pressure plus heat welds them together. This is most commonly used on sheet metal parts and is called spot welding.

FUSION WELDING JOINTS

Some common types of joints used in fusion welding have already been studied. The symbols that represent them in a blueprint are shown in Figure 28.1. These symbols were developed by the American Welding Society (AWS) to help standardize blueprints for easier use on jobs.

Welding information is given on a drawing by placing the symbols in the proper position. (See Figure 28.2.) Keep in mind that when the symbol showing the type of weld is placed above the reference line, the weld is on the side opposite the arrowhead. When it is below the reference line, the weld is on the same side as the arrowhead. When it appears on both sides of the reference line, the weld is on both sides of the joint.

Weld symbols are placed on the reference line with the vertical leg always on the left as you view the symbol. (See Figure 28.3.)

RESISTANCE WELD JOINTS

There are two types of resistance welding (see Figure 28.4):

• Spot welds

• Seam welds

Note how a typical spot weld is placed on the drawing, then take a look at the typical placement of a seam-weld symbol. (See Figure 28.5.) The projection weld is dimensioned on the blueprint as shown in Figure 28.6. Typical flash and upset welds are shown in Figure 28.7. The letter is used to denote that the bulge is to be removed by machining. FW, of course, refers to flash weld, and UW stands for upset weld.

By learning the symbols and how they are used, the job of reading a blueprint becomes much easier.

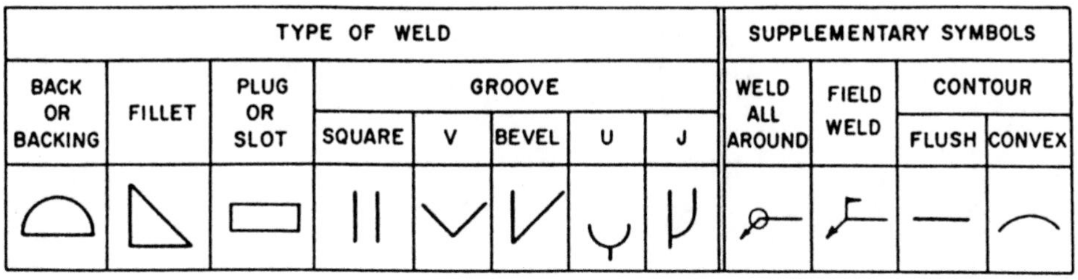

TYPE OF WELD								SUPPLEMENTARY SYMBOLS			
BACK OR BACKING	FILLET	PLUG OR SLOT	GROOVE					WELD ALL AROUND	FIELD WELD	CONTOUR	
			SQUARE	V	BEVEL	U	J			FLUSH	CONVEX
⌒	◺	▭	‖	⋁	⋁	⋃	⋃	⌀	⌐	—	⌒

FIGURE 28.1 Symbols used for fusion welding.

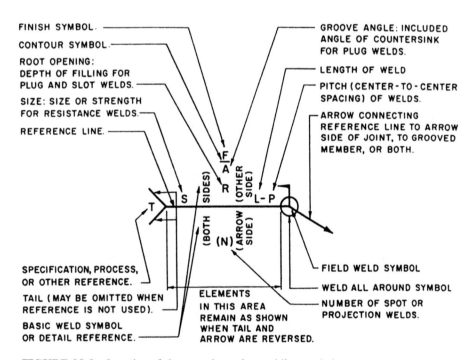

FIGURE 28.2 Location of elements shown in a welding symbol.

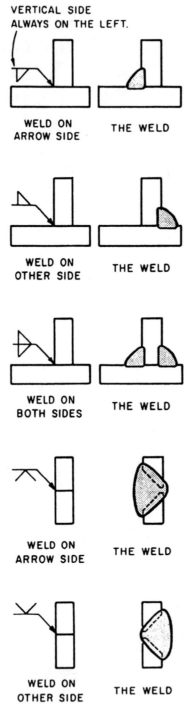

FIGURE 28.3 Placement of the arrowhead on a drawing.

TYPE OF WELD				SUPPLEMENTARY SYMBOLS			
RESISTANCE-SPOT	PROJECTION	RESISTANCE-SEAM	FLASH OR UPSET	WELD ALL AROUND	FIELD WELD	CONTOUR	
						FLUSH	CONVEX
⟨RSW	RPW⟩	RSEW⟩	⊨⟨FW	⟨○	𝄐	—	⌒

FIGURE 28.4 Resistance weld symbols.

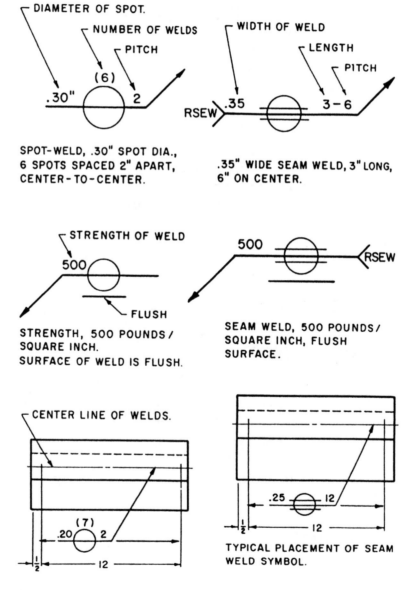

SPOT-WELD, .30" SPOT DIA.,
6 SPOTS SPACED 2" APART,
CENTER-TO-CENTER.

.35" WIDE SEAM WELD, 3" LONG,
6" ON CENTER.

STRENGTH, 500 POUNDS/
SQUARE INCH.
SURFACE OF WELD IS FLUSH.

SEAM WELD, 500 POUNDS/
SQUARE INCH, FLUSH
SURFACE.

TYPICAL PLACEMENT OF SEAM
WELD SYMBOL.

FIGURE 28.5 Placement of symbols on a drawing.

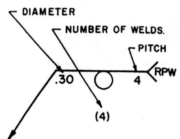

PROJECTION WELD, .30" DIA.,
4 WELDS, 4" ON CENTER,
DIMPLE ON ARROW SIDE OF
PART.

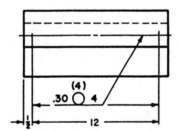

PROJECTION WELD, STRENGTH
400 POUNDS/SQUARE INCH,
FLUSH, DIMPLE ON OTHER SIDE.

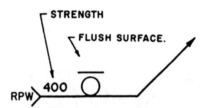

TYPICAL PLACEMENT, OF THE
PROJECTION WELD SYMBOL.

FIGURE 28.6 Projection welds
dimensioning.

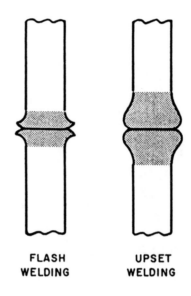

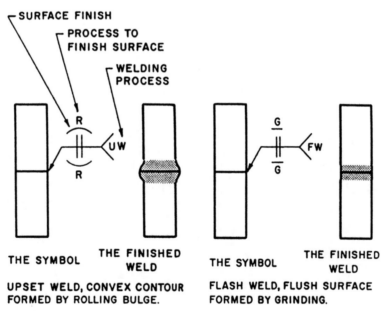

FIGURE 28.7 Flash and upset welds dimensioned.

MULTIPLE-CHOICE EXAM

1.   Today most "blueprints" are black and white and they are referred to as:

      a. diazo prints   **b.** vandyke prints

      c. white prints   **d.** blueprints

      e. all of these   **f.** none of these

2.   Never use the original drawing for reference; you should use a _____ for your shop or on the job location.

      a. copy   **b.** picture

      c. tracing   **d.** written description

3.   Blueprints use a standardized source for their _____.

      a. symbols   **b.** color

      c. prints   **d.** storage

4.   Most drawings are made with _____ in the original rendition.

      a. drafting machines   **b.** felt-tip markers

      c. ink   **d.** pencil

5.   What does CAD stand for?

      a. computer-assisted drafting   **b.** computer-aided drawing

      c. computer-arranged drawings   **d.** none of the above

6.   A _____ drawing is easier to draw than a detailed presentation.

      a. schematic   **b.** sketch

      c. penciled   **d.** blue

7.   As a working drawing for use in the shop or on the job it is easier to use the _____ thread drawing than a detailed presentation.

      a. diazoed   **b.** simplified

      c. ultraviolet   **d.** none of the above

8. The two basic welding processes are fusion and _____.

    a. spot   **b.** torch

    c. seam   **d.** gas

9. Two basic types of electrical resistance welding are spot and _____.

    a. fillet   **b.** seam

    c. square   **d.** plug

10. Welding information is given on a drawing by placing it in the proper _____ on the symbol.

    a. perspective   **b.** manner

    c. location   **d.** direction

TRUE-FALSE EXAM

1. The two basic welding processes are fusion welding and resistance welding.
True   False

2. There are two types of resistance welding, spot and seam.
True   False

3. Standardized welding symbols were developed by Lincoln Electric Company.
True   False

4. It doesn't matter that a symbol is flagged anywhere on the drawing.
True   False

5. The size of a weld is indicated on the welding symbol.
True   False

6. If the length of the weld is the same as the length of the part being welded, no length dimension is given.
True   False

7. The depth of the groove is given to the left of the weld symbol.
True   False

8. When the exposed surface of a resistance spot weld is to be flush, the surface is indicated by adding the flush-contour weld symbol.

    True   False

9. A seam weld is one in which the resistance weld is continuous for its entire length.

    True   False

10. Upset butt welds are made by passing an electric current through two members that are in firm contact with each other.

    True   False

MULTIPLE-CHOICE ANSWERS

1.	E	3.	A	5.	A	7.	B	9.	B
2.	A	4.	D	6.	A	8.	A	10.	C

TRUE-FALSE ANSWERS

1.	T	4.	F	7.	T	9.	T
2.	T	5.	T	8.	T	10.	T
3.	F	6.	T				

Chapter 29

OTHER WELDING METHODS

There are a number of very useful welding processes that use heat sources other than an oxy-fuel gas flame, electrical resistance, or electric arc. Some are quite old; others are among the newest welding processes. Most are suitable for limited but important applications. One such example is thermit welding.

THERMIT WELDING

Thermit welding (TW), an old process, has been replaced by electroslag welding for some purposes. However, it still is very effective. It is widely used for joining thick sections of steel. It works well, particularly where the contour varies, in joining railroad rails or steel castings. It is also very useful in repairing large, broken or cracked steel castings.

Heating and coalescence are produced by superheated molten metal and slag, obtained from the reaction between a metal oxide and aluminum. In addition to furnishing heat, the molten metal also supplies any required filler metal. The thermit process utilizes a mechanical mixture of about one part finely divided aluminum and three parts iron oxide. When the mixture is ignited, it reacts according to the chemical equation:

$$8Al + 3Fe_3O_4 = 9Fe + 4Al_2O_3 + heat$$

This produces a temperature of over 2760 degrees C (5000 degrees F) in about 30 seconds. The ignition temperature of about 1150 degrees C (2100 degrees F) is supplied by a magnesium fuse.

Essential steps for welding by the thermit process are shown in Figure 29.1. Sections to be welded are prepared to provide clearance between them. Wax is used to fill in the gap. The wax is built up to form the desired shape of the weld, riser, and runner system, similar to the procedure used in investment casting. A box is placed around the work and rammed with a material similar to molding sand to form a mold. When the mold is completed, a heating torch is used to dry it and to melt the wax.

Heat is applied until the faces of the work are red. One or more crucibles, filled with the thermit mixture, are set atop the mold; holes in their bottoms connect the runner system in the mold. Crucibles are filled with the mixture and ignited. The molten metal flows out of the crucible. It fills the mold and at the same time supplies sufficient heat to raise the surfaces of the workpieces to a temperature sufficient to produce coalescence. Once the deposited metal has cooled, the mold is removed. The weld can then be ground to its final shape.

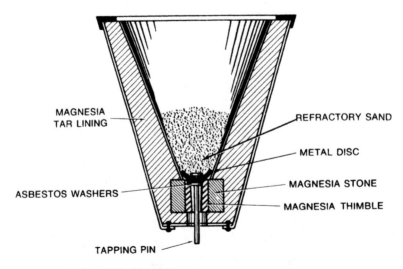

FIGURE 29.1A   Thermit welding.

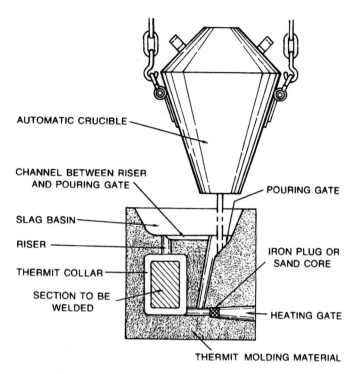

AUTOMATIC CRUCIBLE

CHANNEL BETWEEN RISER
AND POURING GATE

POURING GATE

SLAG BASIN

RISER

IRON PLUG OR
SAND CORE

THERMIT COLLAR

SECTION TO BE
WELDED

HEATING GATE

THERMIT MOLDING MATERIAL

FIGURE 29.1B   Thermit welding.

ELECTROSLAG WELDING

Electroslag welding (ESW) originated in Russia and has been further developed in the United States as well as in several European countries. It is now a very effective process for welding thick sections of steel. Essential components are illustrated in Figure 29.2. Heat is derived from the electrical resistance of the molten slag that is produced between electrode wires in the dip. No arc is needed. The process is different from submerged-arc welding.

Electrical resistance of the material being welded plays no part in producing heat. The temperature of the slag bath is up to 1760 degrees C (3200 degrees F) and is much higher than the melting points of the base metal and the electrodes. That means that the slag melts the edges of the pieces that are being joined and at the same time melts the continuously fed electrodes—this supplies the needed filler metal. In most cases, there is about a 63.5 mm (2 1/2 inch) depth of the molten slag and between 12.7 to 19 mm (1/2 to 3/4 inch) of molten metal confined in the gap between the plates being welded by means of sliding, water-cooled plates. The best conditions for maintaining a deep slag bath exist in vertical joints, so the process is used most often for this type of work. However, it also is used successfully for making circumferential joints in large pipe. It uses special curved slag-holder plates and rotation of the pipe to maintain the area where welding is occurring in a vertical position.

Because very large amounts of weld metal and heat can be supplied, electroslag welding is the best of all the welding processes for making welds in thick plates. Thicknesses up to 457 mm (18 inches) in plain and alloy steels are welded without difficulty. Three electrodes are commonly reciprocated in the direction of plate thickness.

Electroslag welding has virtually replaced thermit welding for welding plates or castings of uniform or uniformly tapering thickness. The choice between electroslag and submerged-arc welding is mostly one of economics. The latter process tends to be more economical for joints below about 63.5 mm (2 1/2 inches), whereas the former is more economical for thicker joints.

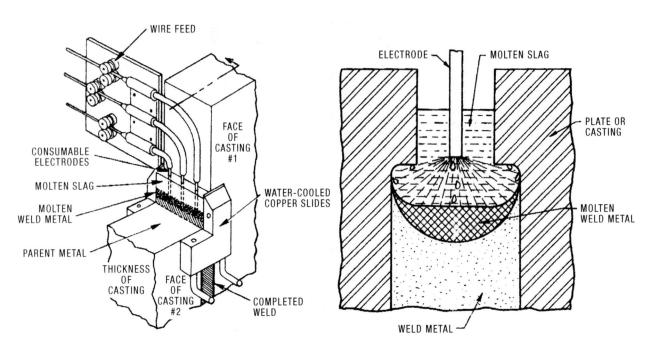

FIGURE 29.2 *(Left)* Arrangement of equipment and work for making a vertical weld by the electro-slag process. *(Right)* Section through work pieces and weld during the making of an electro-slag weld.

FLASH WELDING

In flash welding (FW) the two surfaces that are to be welded are rapidly brought into light contact. This is done while they are connected to an electrical power source. (See Figure 29.3.) As they make contact, a large current passes through the joint and, because the interfaces are not absolutely smooth, some momentary arcing takes place. The resistance to current flow and the incidental arcing produce rapid heating of the metals until they reach welding temperature. After they are forced together with considerable pressure, the entire area of abutment coalesces. This type of action usually is accompanied by some upsetting of the softened metal. The current flow usually is sustained until the entire process is completed. The process involves the following six steps:

- Clamp the work in the machine.
- Apply the welding voltage.

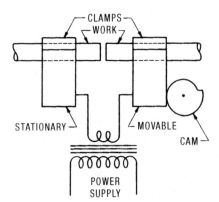

FIGURE 29.3 Principle of the flash-welding process.

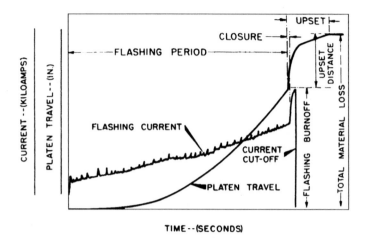

FIGURE 29.4 Relationship of current, platen travel, upset, and time in flash welding.

- Bring the parts into light contact to establish current flow and cause flashing to occur.
- Upset the parts by applying a high pressure when flashing has caused them to reach plastic temperature.
- Cut off the welding current.
- Unclamp the work.

The upsetting action should be sufficient to cause the plastic form of the metal in the abutting surfaces to be displaced outward radially so that all impure metal is squeezed out into the fin, or flash, that is formed. This assures sound metal in the weld. Figure 29.4 depicts the relationship among the current, work (or platen) travel, and time in making a flash weld. Figure 29.5 shows parts before and after being flash-welded. In addition to the six basic steps, preheat and postheat steps can be added in order to improve the metallurgical properties.

Flash welding can be used to butt-weld solid or tubular metals. It is widely used in manufacturing such products as tubular metal furniture, metal windows, and pipe. Except for very small sections, the equipment required is rather large and expensive, but excellent welds can be made at high production rates, making the process well suited to mass production. The resulting flash usually must be removed. However, in most cases no surface preparation is required.

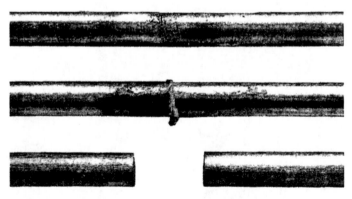

FIGURE 29.5 Flash-welded parts: *(top)* before welding; *(center)* welded, showing flash: *(bottom)* flash ground smooth.

ELECTRON-BEAM WELDING

Electron-beam welding (EBW) is a fusion-welding process in which heat is produced from the impingement of a beam of high-velocity electrons on the metal to be welded. This process was originally developed for obtaining ultra-high-purity welds in reactive and refractory metals in the atomic-energy and rocketry fields. Its unique qualities have led to numerous applicaions.

The electron-optical system used is shown in Figure 29.6. The electrons must be generated and focused in a very high vacuum, and the welding usually is performed at a pressure of 0.13 to 133 mPa. The process can be adopted to weld in pressures of 0.13 to 13 Pa or even at atmospheric pressure, but the penetration of the beam and the depth-to-width ratio are reduced as the pressure increases.

A high-voltage current heats the tungsten filament to about 2204 degrees C (4000 degrees F), causing it to emit high-velocity electrons. By means of a control grid, accelerating anode, and focusing coils, the electrons are converted into a concentrated beam and focused onto the workpiece in a spot from 0.79 to 3.18 mm (1/32 to 1/8 inch) in diameter. Usually the work is enclosed and moved under the electron gun in the vacuum chamber. Under these conditions, the vacuum assures degassification and decontamination of the molten weld metal, and very high-quality welds can be obtained. However, the size of the vacuum chamber required naturally imposes serious limitations on the size of the workpiece that can be accommodated. As a consequence, electron-beam welding machines have been developed that permit the workpiece to remain outside the vacuum chamber. In these machines the electron beam emerges through a small orifice in the vacuum chamber to strike the adjacent workpiece. High-capacity vacuum pumps take care of the leakage through the orifice. Although these machines have some of the advantages of the total-vacuum types, because they do not operate at as low a pressure they do not have as great a penetrating power and, as shown in Figure 29.7, they produce considerably wider welds.

Generally, two ranges of voltages are used in electron-beam welding. High-voltage equipment uses 50 to 100,000 volts. This type provides a smaller spot size and greater penetration than does the lower-voltage

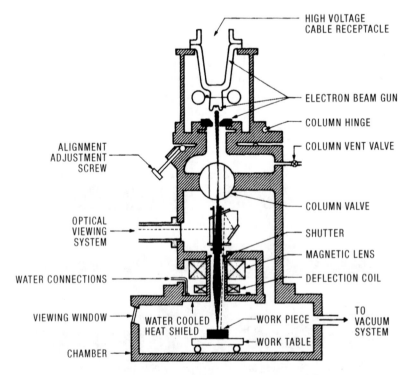

FIGURE 29.6 Schematic diagram of the electron beam welding process.

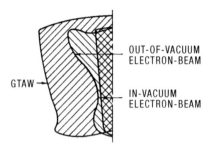

FIGURE 29.7 Comparison of weld widths of GTAW, out-of-vacuum electron beam, and in-vacuum electron beam welds.

type, which uses from 10 to 30,000 volts. However, the high-voltage units, with their high electron velocities, emit considerable quantities of harmful X-rays and require expensive shielding as well as indirect viewing systems for observing the work. X-rays are produced by the low-voltage types too but are sufficiently soft to be absorbed by the walls of the vacuum chamber, and they have adequate spot concentration and penetration for most applications. They also are less critical in adjustment. With them the work can be observed directly through viewing ports.

Materials that are difficult to weld by other processes, such as zirconium, beryllium, and tungsten, can be welded successfully by electron-beam welding, but the weld configuration should be simple and preferably flat. (See Figure 29.8.) Narrow welds can be obtained as well as remarkable penetrations, even through non-adjacent sections of metal. The equipment is quite expensive, and highly skilled labor is required. The process is used mostly where extremely high-quality welds are required or where other welding processes will not produce the required results. However, its unique capabilities have resulted in its routine use in a large number of applications.

ULTRASONIC WELDING

Ultrasonic welding (USW) is a solid-state welding method in which coalescence is produced by localized application of high-frequency vibratory energy channeled to the workpieces as they are held together under pressure. The basic components of the process are shown in Figure 29.9. There is some increase

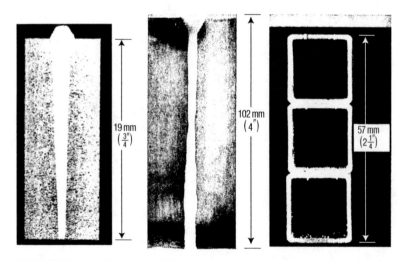

FIGURE 29.8 *(Left to right)* Electron beam welds in 7079 aluminum, thick stainless steel, and a multiple-tier weld in stainless steel tubing.

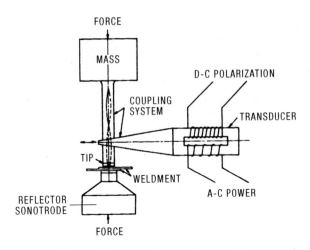

FIGURE 29.9 Schematic diagram of the equipment used
in ultrasonic welding.

in temperature at the faying surfaces, but it is always far below the melting points of the materials. It appears that the rapid reversals of stress along the bond interface play an important role in facilitating coalescence by breaking up and dispersing the mating-surface films.

The ultrasonic transducer used is essentially the same as that used in ultrasonic machining. It is coupled to a force-sensitive system that contains a welding tip on one end. The pieces to be welded are placed between this tip and a reflecting anvil, thereby concentrating the vibratory energy within the work. Either stationary tips, for spot welds, or rotating disks, for seam welds, can be used.

Ultrasonic welding is restricted to joining thin materials:

• Sheet

• Foil

• Wire

The maximum thickness is about 2.54 mm (0.1 inch) for aluminum and 1.02 mm (0.04 inch) for harder metals. However, as indicated in Table 29.1, it is particularly valuable in that numerous dissimilar metals

TABLE 29.1 Metal combinations weld-able by ultrasonic welding.

Metal	Aluminum	Copper	Germanium	Gold	Molybdenum	Nickel	Platinum	Silicon	Steel	Zirconium
Aluminum	X	X	X	X	X	X	X	X	X	X
Copper		X		X		X	X		X	X
Germanium			X	X		X	X	X		
Gold				X		X	X	X		
Molybdenum					X	X			X	X
Nickel						X	X		X	X
Platinum							X		X	
Silicon										
Steel									X	X
Zirconium										X

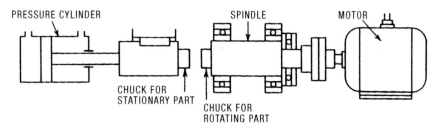

FIGURE 29.10 Arrangement of equipment for inertia (friction) welding.

can readily be joined by the process. Because the temperatures involved are low and no current flow or arcing is involved, intermetallic compounds seldom are formed, and there is no contamination of surrounding areas. The equipment is simple and reliable, and only moderate skill is required of the operator. However, it usually is not economically competitive when other processes can be used.

FRICTION WELDING

The heat for friction welding (FRW)—sometimes called inertia welding—is the result of mechanical friction between two abutting pieces of metal that are held together while one rotates and the other is held stationary. Two basic procedures are used. In the one shown in Figure 29.10, the moving part is held and rotated in a motor-driven collet while the stationary part is pressed against it with sufficient pressure so that the friction quickly generates enough heat to raise the abutting surfaces to welding temperature. As soon as the welding temperature is reached, rotation is stopped and the pressure is maintained or increased until the weld is completed.

In the second process, as indicated in Figure 29.11, one workpiece is gripped in a rotating flywheel, with the kinetic energy of the flywheel being converted into heat by pressing the two workpieces together when the flywheel has attained the desired velocity. Figure 29.12 shows an example of the equipment used and the relationship between surface velocity, torque, and upset.

In both procedures the total cycle time for a weld usually is less than 25 seconds, whereas the actual time for heating and welding is about 2 seconds. No material is melted. Because of the very short period of heating-and thus the lack of time for the heat to disperse from the joint—the weld and heat-affected zones are

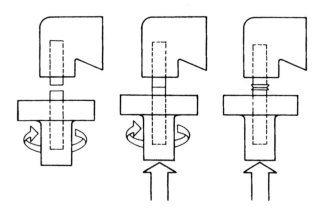

FIGURE 29.11 Schematic representation of the three steps in inertia welding.

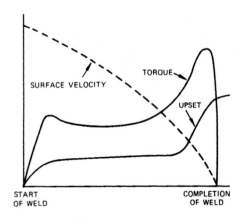

FIGURE 29.12 Inertia-type friction-welding machine and welded part. Relationship between surface velocity, torque, upset, and time in friction welding.

very narrow. Surface impurities are displaced radially into a small flash that can be removed after welding if desired. Because virtually all the energy used is converted into heat, the process is very efficient. However, its use is restricted to joining round bars and tubes of the same size, or to joining bars and tubes to flat surfaces. The ends of the workpieces must be cut true and fairly smooth. It is used primarily for joining pieces of the same metal or plastic, but some dissimilar metals, such as aluminum and magnesium, also can be joined successfully. This process is used extensively in placing childproof plastic caps onto medicine bottles.

LASER-BEAM WELDING

The heat source in laser-beam welding (LBW) is a focused laser beam, usually providing power intensities in excess of 10 kilowatts per square centimeter but with low heat input—0.1 to 10 joules. The high-intensity beam produces a very thin column of vaporized metal extending into the base metal. The column of vaporized metal is surrounded by a liquid pool, which moves along as welding progresses, resulting in welds with depth-to-width ratios greater than 4:1. Laser-beam welds are most effective for autogenous welds, but filler metal can be added. Deep-penetration welds produced by lasers are similar to electron-beam welds, but LBW has several advantages:

- A vacuum environment is not required.

- No X-rays are generated

- The laser beam is easily shaped and directed with reflective optics.

- Because only a light beam is involved, there does not need to be any physical contact between the workpieces and the welding equipment, and the beam will pass through transparent materials, permitting welds to be made inside transparent containers.

Because laser beams are concentrated, laser welds are small—usually less than 0.025 mm (0.001 inch). They are very useful in the electronics industry in connecting leads on small electronic components and in integrated circuitry. Lap, butt, T-, and cross-wire configurations are used. It also is possible to weld lead wires with polyurethane insulation without removing the insulation. The laser evaporates the insulation and completes the weld. Figure 29.13 shows examples of laser welds. Equipment required for laser-beam welding is high in cost but usually is designed for use by semiskilled workers. Its use is restricted to specialized applications.

FIGURE 29.13   Typical welds made by laser welding.

EXPLOSION WELDING

Explosion welding (EXW) is used almost exclusively for bonding sheets of corrosion-resistant metals to heavier plates of base metals—in other words, cladding. It is particularly used for large areas. An explosive material, usually in the form of a sheet, is placed on top of the two layers of metal and detonated progressively. A compressive stress wave, on the order of thousands of megapascals, progresses across the surface of the plates so that a small open angle is formed between the two colliding surfaces. Surface films are liquefied or scarfed off the surfaces and jetted out of the interface, leaving clean surfaces that coalesce under the high pressure. The result is a cold weld with a wavy configuration at the interface where it is heated quickly to upsetting temperature. This is so that the surrounding metal remains relatively cool. Cool water then is sprayed on the plate, and the contraction of the upset spot straightens the buckle over a considerable area. To remove large buckles, it is necessary to repeat the process on several spots.

Flame straightening cannot be used on thin material; the metal adjacent to the heated area must have sufficient rigidity to resist transferring the buckle from one area to another.

WELDING OF PLASTICS

Plastics of the thermoplastic type can be welded successfully using either a hot-gas torch of the type shown in Figure 29.14 or an electrically heated tool.

In the hot-gas torch, a gas—usually air—is heated by an electrical coil as it passes through the torch. Electrical tools similar to an electric soldering iron are moved in contact with the material until the desired temperature is achieved. The heating is localized at 246 to 357 degrees C (475 to 675 degrees F) until the plastic softens. Some pressure then is applied to produce coalescence.

V-groove butt or fillet welds are usually employed. Because the plastic cannot be made to flow, as in fusion welding of metals, some filler material usually has to be added. This is done by using a plastic filler rod that is heated at the same time as the workpieces and then pressed or stretched into the joint. This supplies some of the pressure needed to complete coalescence. (See Figure.29.14.)

Heated-tool welding usually is employed for making lap-seam welds in flexible plastic sheets. Pressure is applied by a roller or other pressure device after the material has been heated.

Butt welds sometimes are made in plastic pipe and rods by friction welding.

FIGURE 29.14 Using a hot-gas torch to make a
weld in plastic pipe.

MULTIPLE-CHOICE EXAM

1. Thermit welding is a(n) _____ welding process.
   a. new
   b. old
   c. contemporary
   d. none of these

2. In thermit welding heating and _____ are produced by superheated molten metal and slag.
   a. coalescence
   b. smoke
   c. oxides
   d. rust

3. Where did Electroslag welding originate?
   a. United States
   b. United Kingdom
   c. China
   d. Russia

4. Electroslag welding is used in making _____ joints.
   a. horizontal
   b. vertical
   c. rough
   d. smooth

5. The choice between electro-lag and submerged-arc welding is mainly one of _____.
   a. economics
   b. coalescence
   c. arc maintenance
   d. none of these

6. Which of the following does not work well with flash welding?
   a. iron
   b. steel
   c. plastics
   d. wood

7. Flash welding relies on _____ to current flow to weld.
   a. resistance
   b. ohms
   c. slag
   d. conduction

8. Electron-beam welding is a _____ -welding process.
   a. friction
   b. fusion
   c. fashionable
   d. none of these

9. Ultrasonic welding is a _____ -welding process.

    a. hot liquid   **b.** solid-state

    c. resistance   **d.** none of these

10. The heat for friction welding is sometimes called:

    a. inertia welding   **b.** quick welding

    c. slow but sure welding   **d.** none of these

11. Which is not an advantage that laser-beam welding has over electron-beam welding?

    a. less power is consumed   **b.** more power is consumed

    c. no X-rays are generated   **d.** none of these

12. Because laser beams are concentrated, laser welds are:

    a. large   **b.** small

    c. of medium-sized   **d.** none of these

13. Explosion welding is used to bond sheets of metal to large, heavy pieces of metal. This process is called:

    a. implanting   **b.** coating

    c. cladding   **d.** none of these

14. Thermoplastics can be welded with a:

    a. hot-gas torch   **b.** gas-heated tool

    c. compressed-air tool   **d.** none of these

TRUE-FALSE EXAM

1. There are a number of very useful welding processes that use heat sources other than oxy-fuel gas flame.

    True   False

2. Thermit welding is a new process of welding.

    True   False

3. Heating and coalescence are produced by superheated molten metal and slag in thermit welding.

True   False

4. Electroslag welding is now a very effective process for welding thick sections of steel.

True   False

5. Because very large amounts of weld metal and heat can be supplied, electroslag welding is the best of the welding processes for making welds in thick plates.

True   False

6. Flash welding is done while two surfaces make contact and a large current flows momentarily through them, causing welding temperatures almost immediately.

True   False

7. Flash welding can be done without large amounts of heat.

True   False

8. Flash welding can be used to butt-weld solid or tubular metals.

True   False

9. Electron-beam welding is a fusion-welding process that uses X-rays to melt metal.

True   False

10. Voltages as high as 100,000 are used in electron-beam welding.

True   False

11. Ultrasonic welding is a solid-state welding method.

True   False

12. Ultrasonic welding is restricted to joining thick materials.

True   False

13. Ultrasonic welding uses vibrations to do the welding.

True   False

14. Friction welding is sometimes called inertia welding.

True   False

15. The friction-welding process is very efficient.

True   False

16. One of the advantages of laser-beam welding is that a vacuum environment is required.

    True   False

17. Explosion welding is used almost exclusively for bonding sheets of corrosion-resistant metals to heavier plates of base metals.

    True   False

18. Another name for explosion welding is cladding.

    True   False

19. V-groove butt or fillet welds are usually employed in plastics welding.

    True   False

20. Butt welds are sometimes made in plastic pipe and rods by friction welding.

    True   False

MULTIPLE-CHOICE ANSWERS

1.	B	4.	B	7.	A	10.	A	13.	C
2.	A	5.	A	8.	B	11.	A	14.	A
3.	D	6.	D	9.	B	12.	B		

TRUE-FALSE ANSWERS

1.	T	6.	T	11.	T	16.	F
2.	F	7.	F	12.	F	17.	T
3.	T	8.	T	13.	T	18.	T
4.	T	9.	T	14.	T	19.	T
5.	T	10.	T	15.	T	20.	T

—NOTES—

Chapter 30

SUBMERGED-ARC WELDING

ubmerged-arc welding is an automatic (or semiautomatic) welding process. It uses a granular flux blanket to completely cover the weld area, while a consumable electrode is continuously and mechanically fed into the arc. (See Figure 30.1.) The name comes from the fact that the arc is hidden or submerged beneath the flux blanket. It is not visible to the welder. This welding process is also referred to as:

- Flux-covered arc welding

- Hidden-arc welding

- Sub-arc welding

- Submerged-melt welding

Union-melt welding is a submerged-arc-welding process developed by the Linde Air Products Company. Unionmelt® is the trade name for the company's specially developed flux. This is an electric-furnace product carefully controlled as to composition and preparation. Linde is a subsidiary of Union Carbide.

Physical characteristics of welds made by the submerged-arc welding process are such that all types of inspection requirements can be met if the procedures specified are followed. Inspections include:

- Full-tensile

- Reduced-tensile

- Free bends

- Nick bends

- X-rays

Basic principles of this welding process center on the concept of a gaseous shield around the arc. Slag over the weld protects both the arc and the molten metal from atmospheric contaminants.

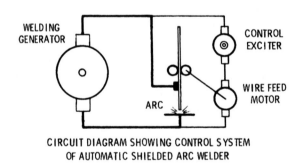

CIRCUIT DIAGRAM SHOWING CONTROL SYSTEM
OF AUTOMATIC SHIELDED ARC WELDER

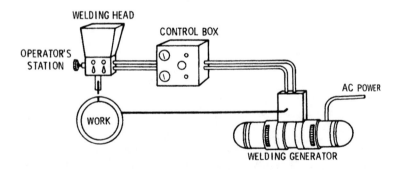

FIGURE 30.1   Submerged welding process.

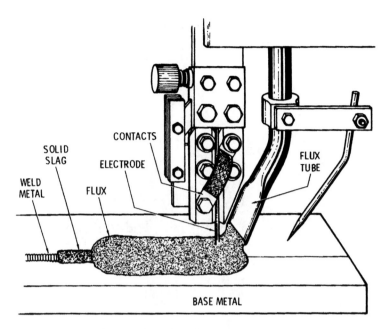

FIGURE 30.2 Distribution of the flux.

Results of this double shielding include deeper penetration, higher welding currents, faster welding, higher weld quality, and lower welding costs.

Instead of flux-coated electrodes, granular flux is automatically fed in the latest development; it is in fact an automatic shielded-metal-arc welding process.

Granular flux is deposited on the joint to be welded and fed deep enough to cover the completed weld. A bare-metal welding-electrode wire is fed into the blanket of flux. The rate of feed is controlled automatically for proper arc length. Direct current (DC) produces the arc between the electrode wire and the joint. The resultant heat from the arc fuses electrode and base metal, producing the weld.

Flux adjacent to the arc melts, floats on the surface of the molten metal, then solidifies as a slag on top of the weld. Since the arc and molten metal are blanketed by flux at all times, the metal is completely protected from contact with the air, assuming maximum quality of welds and making possible the use of very high amperage for faster welding. (See Figure 30.2.)

SUBMERGED-ARC-WELDING EQUIPMENT

A submerged-arc-welding system generally consists of the following basic components:

1. An AC or DC power source

2. A welding rod

3. An electrode reel and motor to feed the electrode wire into the arc

4. A flux hopper and a unit to recover un-fused flux

5. A control unit

Almost any kind of standard DC generator can be used for submerged-arc welding. The heavy-duty AC transformers used in this welding process usually have a 1000-amp capacity (1000 amps for 1 hour of operation, 750 amps of continuous operation). However, both smaller and larger units are also used. Current

FIGURE 30.3   Welding equipment.

higher than that obtainable from one transformer may be had by connecting two or more units in parallel. DC generators may also be connected in parallel to obtain sufficient amperage. (See Figure 30.3.)

Both AC and DC power sources must use a means of remote control for the welding current so that welding can be started and stopped by a switch mounted near the welding head. This is provided by the control unit. If AC is used, a magnetically operated contactor or heavy-duty circuit breaker should be installed in the primary supply of the transformer. The supply leads should have automatic cutout protection of sufficient capacity to protect the primary supply The control unit also indicates voltage and amperage levels and provides controls for current adjustments, the rate of electrode speed, and travel-speed adjustments. (See Figure 30.3.)

WELDING CURRENT

Either DC or single-phase AC may be used for submerged-arc welding. AC is the more frequently used. An open-circuit welding voltage of 60 to 100 volts can be used. Table 30.1 gives approximate maximum currents required for various thicknesses of one-pass for Unionmelt® butt welds and fillet welds.

Sufficient adjustable reactance or resistance should be provided, if using direct current, to permit continuous stepless adjustment and control of the amperage-age during the welding operation. The reactance (or resistance in DC circuits) may be built into the power-supply unit or be connected separately. Actual voltage at the welding zone will be adjusted through the special voltage control at approximately 25 to 50 volts, depending on the size and shape of the weld.

TABLE 30.1 Current ranges for submerged-arc welding electrodes.

Wire diameter (in.)	Current (amp)
5/64	200–600
3/32	230–700
1/8	300–900
5/32	420–1000
3/16	480–1100
7/32	600–1200
1/4	700–1600
5/16	1000–2500
3/8	1500–4000

Submerged-arc welding uses a higher current than that used in the manual arc-welding processes. This results in such welding characteristics as:

- Reduced weld shrinkage
- Minimum distortion of the welded structure
- Faster filler-metal deposition rate
- Faster welding speed

FLUXES AND ELECTRODES

Fluxes used in submerged-arc welding must not produce large amounts of gas. These fluxes are granular fusible substances, available in a number of different grades. Each grade differs somewhat in chemical composition. The selection of an appropriate flux depends on a number of factors, including:

- Type of metal to be welded
- Welding speed
- Thickness of the metal

Flux is fed around the electrode during the welding process. Use just enough flux to submerge the arc. However, there will be no harmful effect on the weld if the arc occasionally breaks out of the flux.

A special aspect of submerged-arc welding is that alloys can be added to the weld through the flux. A mild-steel electrode in conjunction with a special agglomerated-alloy flux is used to do this. Fluxing and alloy materials are ground and mixed together in an agglomeration process so that neither can be separated from the other. The result is an agglomerated-alloy flux that will ensure stable, uniform welds. A wide variety of fluxes are possible by varying the alloy content. Because of the alloying elements that can be incorporated in the flux, the submerged-arc-welding process can be applied to a wide range of jobs using only a small selection of standard electrodes.

Fluxes used in submerged-arc welding are not classified. Therefore, the welder must follow the manufacturer's recommendations for a particular job. Fluxes used in this process are relatively inexpensive. They should be stored in a dry place. If the flux becomes wet, dry it out by heating it to 500 or 600 degrees F.

Fused fluxes may be ground up and reused if they are mixed with at least 75 percent fresh flux. Grinding costs generally make this practice uneconomical.

TABLE 30.2   Typical mechanical electrode lead for circumferential welding.

AWS classification flux-wire	Tensile strength (100 psi)	Yield strength (1000 psi)	Impact strength, charpy v-notch (ft-lb)		Elongation in 2 inch (5)
			at 0° F	at −20° F	
F60, F70-EH14	74	60	19	—	28
F61, F71-EL12	73	60	30	—	26
F61, F71-EH14	77	61	24	—	28
F62, F72-EL12	71	56	—	61	30
F62, F72-EH14	74	58	—	21	28
F62, F72-EM15K	77	64	—	42	26
F62-EL12	68	57	—	45	28
F70-EL12	78	66	21	18	25
F71-EL12	79	68	25	22	27
F71-EM12K	89	77	30	25	25
F71-EM13K	85	70	22	—	25
F72-EM12K	83	75	—	47	26
F72, EM13K	84	74	—	40	27
F72, EM15K	85	72	—	27	24

Instead of stick electrodes, the submerged-arc-welding process uses coils of bare metal wire that are continuously and mechanically fed into the arc. In this respect, submerged-arc welding strongly resembles MIG welding The electrode wire diameters generally range from 1/5 to 1/4 inch and may be purchased in a number of different coil sizes: 1, 5, 10, 25, and 35 pounds and up.

JOINT PREPARATION

Tables 30.1 through 30.4 illustrate equipment settings for submerged-arc welding. On a curved surface, such as small-diameter-girth welds, the speeds and currents are less than the equivalent seam on a flat, horizontal surface. The reason for this is that the molten flux and steel are very fluid and it is necessary to limit the amount of molten material so that it does not run off the curved surface. To control this tendency to spill off, the current and the speed are reduced and the point of welding is 1/2 to 2 1/2 inches off the vertical centerline in the direction opposite the rotation of the work. (See Figure 30.4.)

Where no gap is specified, the seam should be fitted tightly together. In butt seams, if the gap is 1/32 inch or more due to poor fit-up, the seam must be sealed with a sealing bead.

TABLE 30.3   Recommended electrode for circumferential welding.

Wire diameter (in.)	Electrode displacement ahead of vertical center (in.)
1–3	3/8–3/4
3–18	3/4–1
18–36	1 1/4–1 1/2
36–42	1 1/2–1 3/4
42–48	1 3/4–2
48–72	2–2 1/2
72	3

TABLE 30.4 Typical melt-off rates (lb/hr) for EL12 class submerged-arc electrodes.

Welding current (amp)	3/32 inch electrode				5/32 inch electrode				7/32 inch electrode			
	1 inch stickout		3 1/4 inch stickout		1 inch stickout		5 inch stickout		1 inch stickout		5 inch stickout	
	DC(+)	DC(−)	DC(+)	DC(−)	DC(+)	DC(−)	DC(+)	DC(−)	DC(+)	DC(−)	DC(+)	DC(−)
300		6.6	10.5	10.0	15.8							
400		9.5	15.0	16.0	23.2	7.9	13.0	11.9	18.0			
500	12.9	19.5	22.5	32.4	11.2	17.5	19.7	24.4				
600	17.3				14.7	21.8	22.5	31.4	13.1	19.0	18.7	27.0
700					18.3	26.1	30.6	38.9	16.2	23.0	23.7	32.5
800					22.0	31.0	39.1	47.1	19.5	26.5	28.8	38.4
900					26.2				22.9	30.0	34.5	44.7
1000									26.4	34.5	40.0	50.7
1100									32.2	38.5	46.5	57.5
1200									34.8			
1300									40.1			

SPECIAL SUBMERGED-ARC-WELDING PROCESSES

Special jobs or application call for any of several variations of the submerged-arc-welding process. They have been developed to answer the needs of these special applications. For the most part, these variations are multiple-arc-welding procedures (Figure 30.5), including:

- Union-melt arc welding
- Twin-arc welding
- Tandem-arc welding
- Three-o'clock welding
- Series-arc welding
- Arc-strip welding

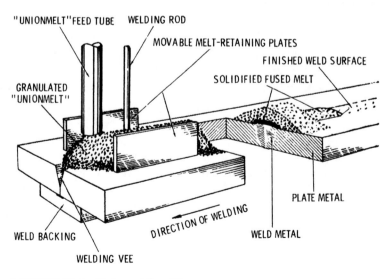

FIGURE 30.4 Unionmelt® welding process.

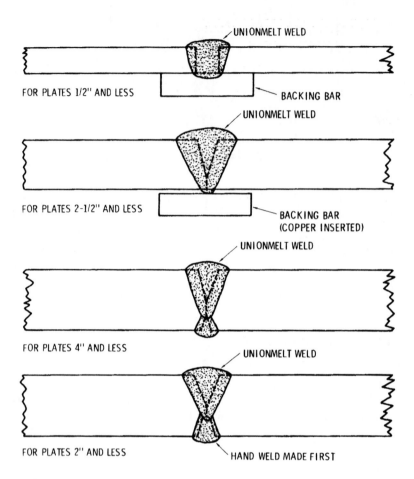

FIGURE 30.5   Edge reparations for Unionmelt welds.

Union-melt welding we have discussed above. Twin-arc welding is a submerged-arc-welding process that consists of two electrodes fed through the same welding head for a fast single-pass method of welding large fillet welds in the flat position and in wide V-joints. DC current is recommended.

Tandem-arc welding employs two or more electrodes used in tandem (one following the other). Each will have its own separate power source. Either AC or DC can be used, often in combination.

Three-o'clock welding (developed by the Lincoln Electric Company) is a submerged-arc-welding process in which the joint can be placed in the horizontal position without loss of flux or weld metal. The electrode is directed into the joint.

Series-arc welding (developed by Union Carbide and Carbon Research Labs) involves the use of two converging electrodes connected in series with a suitable power source for the purpose of making welds with a very shallow penetration. Arc-strip welding (developed by the Arcos Corporation) is a submerged-welding process in which a flat stainless-steel strip 2-inches wide is substituted for the bare-metal wire used by other processes.

SUBMERGED-ARC-WELDING APPLICATIONS

The submerged-arc-welding process is especially suitable for welding soft metals such as:

- Copper and copper alloys
- Nickel and nickel alloys
- Low- and mild-alloy steels
- Stainless steel

This welding process is done in the flat and horizontal positions, which limits its application. Submerged-arc welding is basically an automatic process in which either the equipment or the work may be moved during the welding.

Submerged-arc welding is adaptable to both automatic and semiautomatic welding methods. Both the direction and travel speed of the welding head are controlled by hand. Both procedures can be applied to a vast number of welding jobs ranging from the relatively minor operations of welding axles and spindles to such large-scale ones as welding girders in major construction projects. Enough accessory equipment is available in submerged-arc welding to make it an extremely flexible and adaptable welding method. Submerged-arc welding has become almost synonymous with automatic and semi-automatic welding.

MULTIPLE-CHOICE EXAM

1. Submerged-arc welding is an automatic welding process that uses a _____ -flux blanket to completely cover the weld area while a consumable electrode is continuously and mechanically fed into the arc.

    a. wet   **b.** liquid

    c. dry   **d.** granular

2. Basic principles of submerged-arc welding center on the concept of a _____ shield around the arc.

    a. wooden   **b.** metal

    c. gaseous   **d.** none of the above

3. The rate of wire feed is controlled _____ for proper arc length.

    a. automatically   **b.** manually

    c. by the arc itself   **d.** none of the above

4. Flux adjacent to the arc floats on the surface of the molten metal and solidifies as _____ on top of the weld.

    a. glass   **b.** slag

    c. flux   **d.** none of the above

5. Submerged-arc welding uses a higher _____ than that used in the manual arc-welding processes.

    a. current   **b.** voltage

    c. resistance   **d.** reactance

6. Which of the following is not a characteristic of submerged-arc welding?

    a. reduced weld shrinkage

    b. minimum distortion of the welded structure

    c. a faster welding speed

    d. an easily broken weld joint

7. Selection of an appropriate flux depends on which of these factors?

    a. type of metal to be welded

    b. weather to which joint is to be subjected

    c. age of the metal being welded

    d. none of the above

8. Fluxes used in submerged-arc welding are relatively _____.

    a. inexpensive   **b.** expensive

    c. rare   **d.** common

9. Instead of stick electrodes, the submerged-arc-welding process uses coils of _____ wire.

    a. coated   **b.** bare

    c. kinked   **d.** tungsten

10. Submerged-arc welding is adaptable to both _____ and semiautomatic welding methods.

    a. automatic   **b.** manual

    c. robotic   **d.** none of the above

TRUE-FALSE EXAM

1. Submerged-arc welding is an automatic welding process that uses a granular-flux blanket to completely cover the weld area while a consumable electrode is continuously and mechanically fed into the arc.

    True   False

2. Basic principles of submerged-arc welding center on the concept of a metal shield around the arc.

    True   False

3. The rate of wire feed is controlled manually for proper arc length.

    True   False

4. Flux adjacent to the arc floats on the surface of the molten metal and solidifies as glass on top of the weld.

    True   False

5. Submerged-arc welding uses a higher voltage than that used in manual arc-welding processes.

    True   False

6. Union-melt welding is a submerged-arc process developed by the Linde division of Union Carbide.

    True   False

7. Selection of an appropriate flux depends on the type of metal to be welded.

    True   False

8. Fluxes used in submerged-arc welding are relatively inexpensive.

    True   False

9. Instead of stick electrodes, the submerged-arc-welding process uses coils of bare wire.

    True   False

10. Submerged-arc welding is adaptable to both manual and semiautomatic welding methods.

    True   False

MULTIPLE-CHOICE ANSWERS

1.	D	3.	A	5.	A	7.	A	9.	B
2.	C	4.	B	6.	D	8.	A	10.	A

TRUE-FALSE ANSWERS

1.	T	4.	F	7.	T	9.	T
2.	F	5.	F	8.	T	10.	F
3.	F	6.	T				

Part I

APPENDICES

Welding Licensing Exam Study Guide

—NOTES—

Appendix I
ARC WELDING ELECTRODES

Once it was recognized that an electric arc between two electrodes was a concentrated heat source approaching 7000 degrees F (3871 degrees C)-as early as 1881-various people tried to use the arc. The first was attempts were between a carbon electrode and metal work pieces as the heat source for fusion welding. A simple circuit was used. Just as in gas-flame welding, it was necessary to add a filler metal from a metallic wire. Later, a bare metal wire was used as the electrode. It melted in the arc and automatically supplied the needed filler metal. Results, however, were very uncertain. Because the arc was unstable, a great deal of skill was required to maintain it. Contamination and oxidation of the weld metal resulted from exposure to the atmosphere at such high temperatures. There was little or no understanding of the metallurgical effects and requirements.

The great potential of the process was recognized, particularly as a result of experiences during World War I. However, not much use was made of the process until about 1920, when shielded metal electrodes were developed. They provided a stable arc and shielding from the atmosphere. Some fluxing action for the molten pool of metal solved the major problems in arc welding. Use of the process expanded very rapidly, and, as you know, has continued to do so ever since. Today we have a considerable variety of arc-welding processes available.

All arc-welding processes use essentially the same basic circuit, except that alternating current is used at least as much as direct current. See Figure A1.1. When direct current is used, if the work is made positive (the anode of the circuit) and the electrode is made negative, *straight polarity* is said to be used. When the work is negative and the electrode is positive, the polarity is *reversed.* When bare electrodes are used, greater heat is liberated at the anode. Certain shielded electrodes, however, change the heat conditions and are used with reverse polarity.

All arc welding is done with metal electrodes. In one type the electrode is consumed. That means it supplies the needed filler metal to fill in the voids in the joint and to speed the welding process. In this case the electrode has a melting point below the temperature of the arc. Small droplets are melted from the end of the electrode and pass to the parent metal. The size of these droplets varies greatly, and the mechanism of the transfer varies with different types of electrodes and processes. As the electrode melts, the arc length and the resistance of the arc path vary as results of the change in arc length and the presence or absence of metal particles in the path. These changes require that the electrode be moved toward or away from the work to maintain the arc and satisfactory welding conditions.

Today, no ordinary manual arc welding is done with bare electrodes. Shielded or covered electrodes are always used. In a large amount of automatic and semiautomatic arc welding, the electrode is a continuous, bare metal wire. However, this type of electrode is always used in conjunction with a separate shielding and arc-stabilizing medium and automatic feed-controlling devices that maintain the proper arc length.

In the other types of metal arc welding, the electrode is made of tungsten, which is not consumed by the arc except by relatively slow vaporization. In some applications a separate filler wire must be used to supply the needed metal.

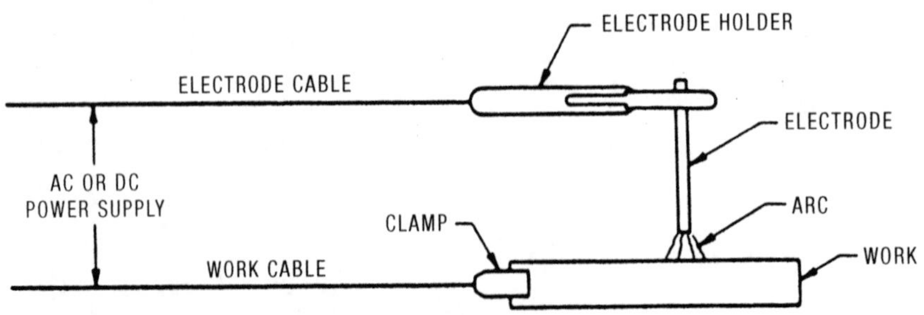

FIGURE A1.1 Basic arc-welding circuit.

Shielded metal arc welding uses electrodes that consist of metal wire, usually from 1.59 to 9.53 mm (1/16 to 3/8 inch) in diameter. An extruded coating containing chemical components add a number of desirable characteristics, including all or a number of the following:

- a protective atmosphere

- arc stability

- flux to remove impurities from the molten metal

- protective slag to accumulate impurities, prevent oxidation, and slow down the cooling of the weld metal

- reduced weld-metal spatter and increased efficiency of deposition

- alloying elements

- arc penetration effects

- influences on the shape of the weld bead

- additional filler metal

COATED ELECTRODES

Coated electrodes are classified on the basis of:

- the tensile strength of the deposited weld metal

- the welding position in which they may be used

- the type of current and polarity they create (if direct current)

- the type of covering

A system of four- or five-digit codes is used to designate types of coated electrodes. For example, type E7016 is a low-alloy steel electrode that will provide a deposit having a minimum tensile strength of 70,000 psi in the non-stress relieved condition. It can be used in all positions, with either alternating current or reverse-polarity direct current, and it has a low-hydrogen-type coating.

In general, the cellulosic coatings contain about 50 percent SiO_2, 10 percent TiO_2, small amounts of FeO, MgO, and Na_2O, and about 30 percent volatile matter. The titaniam coatings have about 30 percent SiO_2, 50 percent TiO_2, small amounts of FeO, MgO, Na_2O, and Al_2O_3, and about 5 percent volatile material.

The low-hydrogen coatings typically contain about 28 percent TiO_2 plus ZrO_2 and 25 percent CaO plus MgO. They eliminate dissolved hydrogen in the deposited weld metal and thus prevent micro-cracking. To be effective they must be baked just prior to use to assure the removal of all moisture from the coating.

Electrodes are marked with colors in accordance with a standard established by the NEMA (National Electrical Manufacturers Association). This coding is intended to make their types readily identifiable.

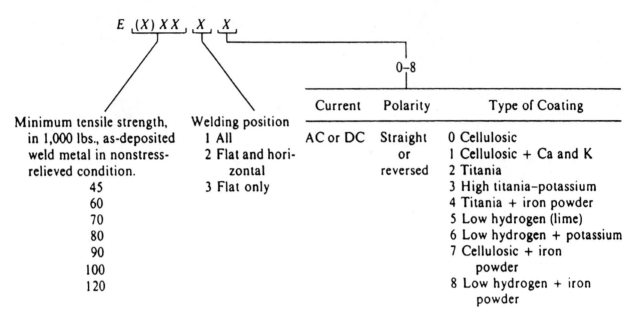

FIGURE A1.2   Designation system for arc-welding electrodes.

COATINGS

When the coating on the electrode melts and vaporizes, it forms a protective atmosphere around the work and the arc. This atmosphere stabilizes the arc and protects the molten and hot metal from contamination. Flux bonds with any impurities in the molten metal and floats them to the surface to be entrapped in the slag coating that forms over the weld. The slag coating protects the cooling metal from oxidation and slows the cooling rate to prevent hardening. Slag is easily chipped from the weld when it has cooled.

Electrodes with iron powder in the coating are used extensively, particularly in production-type welding. They greatly increase the amount of metal that can be deposited with a given size of electrode wire and current, reducing welding costs.

An insulating coating on one type of electrode melts so slowly that it protrudes slightly beyond the melting filler wire. Dragging this type of electrode along the work will maintain the proper arc length. For this reason these electrodes are called *contact* or *drag* electrodes.

A very large amount of welding is still done with ordinary shielded electrodes. In recent years there has been a great increase in the use of other methods of shielding. That is because these methods permit the use of continuous electrodes as well as automatic electrode-feeding devices. Figure A1.2 shows the designation system for arc-welding electrodes.

Appendix 2
CONVERSION TABLES

Metric And English Measures

The metric unit of length is the meter = 39.37 inches.

The metric unit of weight is the gram = 15.432 grains.

The following prefixes are used for subdivisions and multiples: milli = 1/1000; centi = 1/100; deci = 1/10; hecto = 100; kilo = 1000; myria = 10,000.

Equivalent Measures

MEASURES OF LENGTH

Metric	English
1 meter	= 39.37 inches, 3.28083 feet, or 1.09361 yards
0.3048 meter	= 1 foot
1 centimeter	= 0.3937 inch
2.54 centimeters	= 1 inch
1 millimeter	= 0.03937 inch, or nearly 1/25 inch
25.4 millimeters	= 1 inch
1 kilometer	= 1093.61 yards, or 0.62137 mile

The ratio 25.4 mm = 1 inch is used to convert millimeters to inches.

MEASURES OF WEIGHT

Metric	English
1 gram	= 15.432 grains
0.0648 gram	= 1 grain
28.35 grams	= 1 ounce avoirdupois
1 kilogram	= 2.2046 pounds
0.4536 kilogram	= 1 pound

$$\left. \begin{array}{l} \text{1 metric ton} \\ \text{1000 kilograms} \end{array} \right\} = \left\{ \begin{array}{l} \text{0.9842 ton of 2240 pounds} \\ \text{19.68 cwt} \\ \text{2204.6 pounds} \end{array} \right.$$

$$\left. \begin{array}{l} \text{1.016 metric tons} \\ \text{1016 kilograms} \end{array} \right\} = \text{1 ton of 2240 pounds}$$

MEASURES OF VOLUME OR CAPACITY

Metric	English

$$\text{1 liter (= 1 cubic decimeter)} = \left\{ \begin{array}{l} \text{61.023 cubic inches} \\ \text{0.03531 cubic foot} \\ \text{0.2642 gal (American)} \\ \text{2.202 lb of water at } 62°F \end{array} \right.$$

28.317 liters = 1 cubic foot

3.785 liters = 1 gallon (American)

4.543 liters = 1 gallon (Imperial)

English Conversion Factors

Length

Inches	×	0.0833	= feet
Inches	×	0.02778	= yards
Inches	×	0.00001578	= miles
Feet	×	0.3333	= yards
Feet	×	0.0001894	= miles
Yards	×	36.00	= inches
Yards	×	3.00	= feet
Yards	×	0.0005681	= miles
Miles	×	63360.00	= inches
Miles	×	5280.00	= feet
Miles	×	1760.00	= yards
Circumference of circle	×	0.3188	= diameter
Diameter of circle	×	3.1416	= circumference

Area

Square inches	×	0.00694	= square feet
Square inches	×	0.0007716	= square yards
Square feet	×	144.00	= square inches
Square feet	×	0.11111	= square yards
Square yards	×	1296.00	= square inches
Square yards	×	9.00	= square feet
Dia. of circle squared	×	0.7854	= area
Dia. of sphere squared	×	3.1416	= surface

Volume

Cubic inches	×	0.0005787	= cubic feet
Cubic inches	×	0.00002143	= cubic yards
Cubic inches	×	0.004329	= U. S. gallons
Cubic feet	×	1728.00	= cubic inches
Cubic feet	×	0.03704	= cubic yards
Cubic feet	×	7.4805	= U. S. gallons
Cubic yards	×	46656.00	= cubic inches
Cubic yards	×	27.00	= cubic feet
Dia. of sphere cubed	×	0.5236	= volume

Weight

Grains (avoirdupois)	×	0.002286	= ounces
Ounces (avoirdupois)	×	0.0625	= pounds
Ounces (avoirdupois)	×	0.00003125	= tons
Pounds (avoirdupois)	×	16.00	= ounces
Pounds (avoirdupois)	×	0.01	= hundredweight
Pounds (avoirdupois)	×	0.0005	= tons
Tons (avoirdupois)	×	32000.00	= ounces
Tons (avoirdupois)	×	2000.00	= pounds

(continued)

English Conversion Factors *(continued)*

Energy

Horsepower	×	33000.0	= ft lb per min
Btu	×	778.26	= ft lb
Ton of refrigeration	×	200.0	= Btu per min

Pressure

Lb per sq in.	×	2.31	= ft of water (60°F)
Ft of water (60°F)	×	0.433	= lb per sq in.
In. of water (60°F)	×	0.0361	= lb per sq in.
Lb per sq in.	×	27.70	= in. of water (60°F)
Lb per sq in.	×	2.041	= in. of Hg (60°F)
In. of Hg (60°F)	×	0.490	= lb per sq in.

Power

Horsepower	×	746.0	= watts
Watts	×	0.001341	= horsepower
Horsepower	×	42.4	= Btu per min

Water Factors
(at point of greatest density—39.2°F)

Miners inch (of water)	×	8.976	= U. S. gal per min
Cubic inches (of water)	×	0.57798	= ounces
Cubic inches (of water)	×	0.036124	= pounds
Cubic inches (of water)	×	0.004329	= U. S. gallons
Cubic inches (of water)	×	0.003607	= English gallons
Cubic feet (of water)	×	62.425	= pounds
Cubic feet (of water)	×	0.03121	= tons
Cubic feet (of water)	×	7.4805	= U. S. gallons
Cubic feet (of water)	×	6.232	= English gallons
Cubic foot of ice	×	57.2	= pounds
Ounces (of water)	×	1.73	= cubic inches
Pounds (of water)	×	26.68	= cubic inches
Pounds (of water)	×	0.01602	= cubic feet
Pounds (of water)	×	0.1198	= U. S. gallons
Pounds (of water)	×	0.0998	= English gallons
Tons (of water)	×	32.04	= cubic feet
Tons (of water)	×	239.6	= U. S. gallons
Tons (of water)	×	199.6	= English gallons
U. S. gallons	×	231.00	= cubic inches
U. S. gallons	×	0.13368	= cubic feet
U. S. gallons	×	8.345	= pounds
U. S. gallons	×	0.8327	= English gallons
U. S. gallons	×	3.785	= liters
English gallons (Imperial)	×	277.41	= cubic inches
English gallons (Imperial)	×	0.1605	= cubic feet
English gallons (Imperial)	×	10.02	= pounds
English gallons (Imperial)	×	1.201	= U. S. gallons
English gallons (Imperial)	×	4.546	= liters

Metric Conversion Factors

Length

Millimeters	×	0.03937	= inches
Millimeters	÷	25.4	= inches
Centimeters	×	0.3937	= inches
Centimeters	÷	2.54	= inches
Meters	×	39.37	= inches
Meters	×	3.281	= feet
Meters	×	1.0936	= yards
Kilometers	×	0.6214	= miles
Kilometers	÷	1.6093	= miles
Kilometers	×	3280.8	= feet

Area

Sq millimeters	×	0.00155	= sq in.
Sq millimeters	÷	645.2	= sq in.
Sq centimeters	×	0.155	= sq in.
Sq centimeters	÷	6.452	= sq in.
Sq meters	×	10.764	= sq ft
Sq kilometers	×	247.1	= acres
Hectares	×	2.471	= acres

Volume

Cu centimeters	÷	16.387	= cu in.
Cu centimeters	÷	3.69	= fl drs (U.S.P.)
Cu centimeters	÷	29.57	= fl oz (U.S.P.)
Cu meters	×	35.314	= cu ft
Cu meters	×	1.308	= cu yards
Cu meters	×	264.2	= gal (231 cu in.)
Liters	×	61.023	= cu in.
Liters	×	33.82	= fl oz (U.S.P.)
Liters	×	0.2642	= gal (231 cu in.)
Liters	÷	3.785	= gal (231 cu in.)
Liters	÷	28.317	= cu ft
Hectoliters	×	3.531	= cu ft
Hectoliters	×	2.838	= bu (2150.42 cu in.)
Hectoliters	×	0.1308	= cu yd
Hectoliters	×	26.42	= gal (231 cu in.)

(continued)

Metric Conversion Factors *(continued)*

Weight

Grams	×	15.432	= grains
Grams	÷	981.0	= dynes
Grams (water)	÷	29.57	= fl oz
Grams	÷	28.35	= oz avoirdupois
Kilograms	×	2.2046	= lb

Weight

Kilograms	×	35.27	= oz avoirdupois
Kilograms	×	0.0011023	= tons (2000 lb)
Tonneau (metric ton)	×	1.1023	= tons (2000 lb)
Tonneau (metric ton)	×	2204.6	= lb

Unit Weight

Grams per cu cent	×	27.68	= lb per cu in.
Kilo per meter	×	0.672	= lb per ft
Kilo per cu meter	×	0.06243	= lb per cu ft
Kilo per cheval	×	2.235	= lb per hp
Grams per liter	×	0.06243	= lb per cu ft

Pressure

Kilograms per sq cm	×	14.223	= lb per sq in.
Kilograms per sq cm	×	32.843	= ft of water (60°F)
Atmospheres (international)	×	14.696	= lb per sq in.

Energy

Joule	×	0.7376	= ft lb
Kilogram meters	×	7.233	= ft lb

Power

Cheval vapeur	×	0.9863	= hp
Kilowatts	×	1.341	= hp
Watts	÷	746.0	= hp
Watts	×	0.7373	= ft lb per sec

Miscellaneous

Kilogram calorie	×	3.968	= Btu
Standard gravity (sea level 45° lat.)	÷	980.665	= centimeters per sec per sec
Frigories/hr. (French)	÷	3023.9	= tons refrigeration

Weights of Steel and Brass Bars

(Weight of Bar I Foot Long)

STEEL—Weights cover hot worked steel about 0.50 percent carbon. One cubic inch weighs 0.2833 lb. High-speed steel, 10 percent heavier.

BRASS—One cubic inch weighs 0.3074 lb.

Actual weight of stock may be expected to vary somewhat from these figures because of variations in the manufacturing processes.

SIZE Inches	Steel			Brass		
	lb per ft			lb per ft		
	Round	Square	Hex	Round	Square	Hex
1/16	0.0104	0.013	0.0115	0.0113	0.0144	0.0125
1/8	0.042	0.05	0.046	0.045	0.058	0.050
3/16	0.09	0.12	0.10	0.102	0.130	0.112
1/4	0.17	0.21	0.19	0.18	0.23	0.20
5/16	0.26	0.33	0.29	0.28	0.36	0.31
3/8	0.38	0.48	0.42	0.41	0.52	0.45
7/16	0.51	0.65	0.56	0.55	0.71	0.61
1/2	0.67	0.85	0.74	0.72	0.92	0.80
9/16	0.85	1.08	0.94	0.92	1.17	1.01
5/8	1.04	1.33	1.15	1.13	1.44	1.25
11/16	1.27	1.61	1.40	1.37	1.74	1.51
3/4	1.50	1.92	1.66	1.63	2.07	1.80
13/16	1.76	2.24	1.94	1.91	2.43	2.11
7/8	2.04	2.60	2.25	2.22	2.82	2.45
15/16	2.35	2.99	2.59	2.55	3.24	2.81
1	2.67	3.40	2.94	2.90	3.69	3.19
1 1/16	3.01	3.84	3.32	3.27	4.16	3.61
1 1/8	3.38	4.30	3.73	3.67	4.67	4.04
1 3/16	3.77	4.80	4.16	4.08	5.20	4.51
1 1/4	4.17	5.31	4.60	4.53	5.76	4.99
1 5/16	4.60	5.86	5.07	4.99	6.35	5.50
1 3/8	5.04	6.43	5.56	5.48	6.97	6.04
1 7/16	5.52	7.03	6.08	5.99	7.62	6.60
1 1/2	6.01	7.65	6.63	6.52	8.30	7.19

Useful Information

To find the side of an equilateral triangle inscribed in a circle, multiply the diameter of the circle by 0.866.

To find the diameter of a circle inscribed in an equilateral triangle, multiply a side of the triangle by 0.57735.

To find the circumference of a circle, multiply the diameter by 3.1416.

To find the diameter of a circle, multiply the circumference by 0.31831.

To find the area of a circle, multiply the square of the diameter by 0.7854.

The radius of a circle $\times$ 6.283185 = the circumference.

The square of the circumference of a circle $\times$ 0.07958 = the area.

Half the circumference of a circle $\times$ half its diameter = the area.

The circumference of a circle $\times$ 0.159155 = the radius.

The square root of the area of a circle $\times$ 0.56419 = the radius.

The square root of the area of a circle $\times$ 1.12838 = the diameter.

To find the diameter of a circle equal in area to a given square, multiply a side of the square by 1.12838.

To find the side of a square equal in area to a given circle, multiply the diameter by 0.8862.

To find the side of a square inscribed in a circle, multiply the diameter by 0.7071.

To find the side of a hexagon inscribed in a circle, multiply the diameter of the circle by 0.500.

To find the diameter of a circle inscribed in a hexagon, multiply a side of the hexagon by 1.7321.

To find the area of the surface of a ball (sphere), multiply the square of the diameter by 3.1416.

To find the volume of a ball (sphere), multiply the cube of the diameter by 0.5236.

Appendix 3 _____

CERTIFICATIONS

Welding Licensing Exam Study Guide

THE AMERICAN WELDING SOCIETY (AWS) is the certification association for welding in the United States. It is located in Miami, Florida and is easily reached on the internet with a simple AWS in the search box of your computer. Prices, schedules, and certifications available by the association are posted on the net.

Certification Programs from AWS

Certified Welding Inspector
Certified Associate Welding Inspector
Senior Certified Welding Inspector
Renewal & Upgrade of the above
9-year Re-Certification
Reciprocity (CWB) for Certified Welding Inspector
Certified Welding Educator
Radiographic Interpreter
Certified Welding Supervisor
Certified Welding Engineer
Certified Robotic Arc Welding
Accredited Test Facility
Certified Welding Fabricator
Certified Welder

Some of the Welding Specialty areas:
Welding Examiner
Welding Inspector
Visual Welding Inspector
Senior Welding Inspector
Structural Welding Certification (Wisconsin)
Underwater Welding
Marine Welding
International Certification

Appendix 4
WELDING TERMINOLOGY

Welding Licensing Exam Study Guide

-A-

abrasion

Wear most often caused by the rubbing together of two or more rough surfaces.

ac

Standard abbreviation for alternating current. The type of electrical power that reverses its direction of flow periodically. Measured in Hertz (Hz).

acetone

Volatile, flammable liquid.

acetylene

Gas composed of two parts carbon and two parts hydrogen (C_2H_2). When burned in an atmosphere of oxygen, it produces one of the highest flame temperatures available. It is a colorless gas with an obnoxious, sweet odor.

acetylene cylinder

Specially built container used to store and ship acetylene. Also, referred to as a tank or bottle.

acetylene regulator

Automatic valve used to reduce acetylene cylinder pressures to working torch pressures. It also keeps the flow of acetylene constant.

acid core solder

See cored solder (most often used term).

activated rosin flux

Rosin- or resin-base flux with an additive to increase wetting action of solder.

alloy

Mixture of two or more metals to produce specific qualities such as hardness and ductility.

annealing

The softening of metals by heat treating. The process most often involves heating the metal to a critical temperature and then cooling it slowly.

anode

Positive (+) terminal of an electrical circuit or battery.

arc

Flow of electricity through a gaseous space or air gap.

arc cutting

Group of cutting processes. The severing or removing of metals accomplished by melting them with heat from an arc between an electrode and the parent metal.

arc-seam weld

Weld bead with an arc welding unit.

arc-spot weld

Spot weld made by the arc welding process.

arc voltage

Electrical potential across an arc. Electrical pressure or voltage of an arc.

arc welding

Fusing metals using the arc welding process.

Austinite

A solid solution of carbon and other elements in iron which forms when the allotropic change takes place. The metal has a face-centered, cubic space lattice.

automatic oxygen cutting

Oxygen cutting with equipment that is fully automated. Operator sets up the work initially equipment does the rest.

automatic welding

Some type of arc welding with all welding operations controlled and initiated by automation.

AWS

American Welding Society.

axis of a weld

Imaginary line along the center of gravity of a weld, perpendicular to a cross section of the weld metal.

- B -

backfire

Momentary retrogression or burning back of the torch flame into the torch tip. Occurs immediately following the withdrawal of the tip from the work. Gases may be reignited by the hot work piece. Otherwise, the use of a lighter may be necessary.

back gouging

Forming of a bevel or groove on the other side of a partially welded joint. Assures penetration upon subsequent welding from that side.

backhand welding

Method of welding in which the torch and rod are so disposed in the vee that the torch flame points back at the completed weld, enveloping the newly deposited metal. Rod is interposed between the torch and the weld.

balling up

Formation of globules of molten brazing filler material or flux by failure to adequately wet or tin the base metal. Also, professional welder's term (slang) used to describe a job that has been done poorly.

base metal

Material composing the pieces to be joined by welding. Also called parent metal.

bead

Denotes the appearance of the finished weld and describes the neatness of the ripples formed by the metal while it was in a semi-liquid state.

bevel

Special preparation of metal to be welded. The edge is ground or cut to an angle other than 90 degrees to the surface of the parent metal.

blind joint

Joint in which no portion is visible.

blowpipe

Oxyacetylene torch handle. No longer used except in experiments.

bond

Junction of the weld metal and the base metal.

braze

Weld wherein coalescence is produced by heating to suitable temperatures above 800 degrees F and by using a nonferrous filler metal having a melting point below that of the base metal. Filler metal is distributed in the joint by capillary attraction.

braze weld

Weld wherein coalescence is produced by heating to a suitable temperature above 800 degrees F by using a nonferrous filler metal having a melting point below that of the base metal. Filler metal is not distributed in the joint by capillary attraction.

Brinell test

Method for determining surface hardness of metallic materials.

bronze weld

See braze weld.

buildup

Amount of weld face or bead that extends above the surface of joined metals.

burned metal

Term usually applied to the metal that has been combined with oxygen so that some of the carbon changed into carbon dioxide (CO_2) and some of the iron into iron oxide (Fe_2O_3).

burning

Violent combination of oxygen with any substance that produces heat. Sometimes used for the term flame cutting.

butt joint

Assembly in which two pieces to be joined are in the same plane, with the edge of one piece touching the edge of the other.
See vee groove.

butt weld

Actual weld in a butt joint.

-C-

capillary action

Property of a liquid to move into small spaces. Ability to "wet" these small surfaces.

calcium carbide

(CaC_2). Chemical compound of calcium and carbon usually prepared by fusing lime and coke in an electric furnace. Compound reacts with water to form acetylene gas.

carbon

Element that, when combined with iron, forms various kinds of steel. Changing carbon content of steel causes changes in its physical properties. Carbon is also used in a solid form as an electrode for arc welding, as a mold to hold weld metal, and for brushes in electrical motors.

carbonizing

See carburizing.

carburizing

Carburizing flame is an oxygen-fuel gas flame. It has a slight excess of fuel gas.

case hardening

Adding carbon to the surface of a mild-steel object and heat treating to produce a hard surface.

castings

Metallic forms that are produced by pouring molten metal into a shaped container or mold.

cathode

Electrical term for a negative (−) terminal.

Celsius

Temperature scale used in the metric system. Zero represents the freezing point of water and 100 is its boiling point (at sea level). Convert to Fahrenheit: multiply by 9, divide by 5, and add 32. A Swede by the name of Celsius invented the centigrade system.

chamfering

See bevel.

coalescence

Growing together or growth into one body of the base metal parts.

coated electrode

Metal rod used in arc welding. Rod has a covering of materials to aid in the arc welding process.

complete joint penetration

Joint penetration extends completely through the joint.

concave fillet weld

Fillet weld with a concave face.

concave root surface

Root surface that is concave.

concave weld face

Weld with the center of its face below the weld edges. Indented weld bead.

cone

Conical part of a gas flame next to the orifice of the tip.

continuous weld

Making the complete weld in one operation.

convex fillet weld

Fillet weld having a convex face.

cored solder

Solder wire or bar containing flux as a core.

corner flange weld

Flange weld with only one member flanged at the location of welding.

corner joint

Junction formed by the edges of two pieces of metal touching each other at an angle of about 90 degrees.

corrosive flux

Flux with a residue that chemically attacks the base metal. May be composed of inorganic salts and acid, organic salts and acids, or activated rosins and resins.

coupons

Specimens cut from the weld assembly for testing purposes.

covered electrode

See coated electrode.

cracking

Action of opening a valve on a tank of fuel gas or oxygen and then closing the valve immediately.

crater

Depression in the face of a weld, usually at the termination of an arc weld.

creep

Gradual increase of the working pressure (as indicated on the gauge) resulting from the regulator seat not closing tightly against the inlet nozzle and permitting the high-pressure gas to leak into the low-pressure chamber. When this condition exists, the regulator should be repaired by a qualified person before use.

crown

Curve or convex surface of a finished weld.

cutting attachment

Device that is attached to a gas welding torch handle to convert it into an oxygen cutting torch.

cutting flame

Cutting by rapid oxidation process at a high temperature. Produced by a gas flame accompanied by a jet action. Jet action blows the oxides away from the cut.

cutting tip

That part of an oxygen cutting torch from which the gases issue and burn.

cylinder

Portable metallic container for storage and transmission of compressed gases.

- D -

dead annealed

The result of heating a work-hardened metal to a red color and immediately quenching in water. This softens the metal and renders it workable again.

deoxidized copper

Copper from which the oxygen has been removed by the addition of a deoxidizer, phosphorus, or silicon. This lowers the electrical conductivity but yields a product more suitable for oxyacetylene welding.

deposited metal

Filler metal that has been added during a welding operation.

depth of fusion

Distance that fusion extends into the base metal or the previous pass from the surface melted during welding.

dip brazing

Brazing process in which the heat required is furnished by a molten chemical or metal bath. When a molten chemical bath is used, the bath may act as a flux. When a molten metal bath is used, the bath provides the filler metal.

dip soldering

Soldering process in which the heat required is furnished by a molten metal bath that provides the solder.

direct polarity

Direct current flowing from anode (base metal) to cathode (electrode). The electrode is negative, and the base metal is positive.

distortion

Warping of a metal or metal surface as a result of uneven cooling.

down-hand welding

Welding in a flat position.

drag

Term used in oxyacetylene cutting. It is the amount by which the oxygen jet falls behind the perpendicular in passing through the material.

drop-thru

Undesirable sagging or surface irregularity, usually encountered when brazing or welding near the *solidus* of the base metal. The condition is caused by over heating with rapid diffusion or alloying between the filler metal and the base metal.

ductility

Property of metals that enables them to be mechanically deformed without breaking when cold.

- E -

edge joint

Welded joint connecting the edges of two or more parallel or nearly parallel parts.

electrode

Substance that brings electricity up to the point where the arc is to be found.

elongation

Total amount of stretching of a specimen, produced in a tensile strength test.

erosion

Reducing the size of or wearing away of an object because of liquid or gas impact.

expansion

Increase in one or more of the dimensions of a body, caused usually by a rise in temperature.

explosion welding

Solid-state welding process wherein coalescence is affected by high-velocity movement produced by a controlled detonation.

- F -

face of weld

Exposed surface of a weld.

Fahrenheit

Temperature scale used in most English-speaking countries where 32 degrees F is the temperature at which water will freeze and 212 degrees F is the temperature at which water will boil, at sea level. Symbol is F.

faying

Process of fitting exactly in place, such as in spot welding having the metal pieces exactly where they are supposed to be.

ferrous metals

Metals and alloys of which the principal base or constituent is iron. These metals are magnetic as well.

filler metal

Material to be added in making a weld.

fillet

Weld metal in the interval vertex, or corner, of the angle formed by two pieces of metal, giving the joint additional strength to withstand unusual stresses.

fillet weld

Metal fused into a corner formed by two pieces of metal whose welded surfaces are approximately 90 degrees to each other.

flame cutting

Cutting performed by an oxygen-fuel gas torch flame that has an oxygen jet.

flanged edge joint

Joint in two pieces of metal formed by flanging the edges at 90 degrees to the plates and joining with an edge weld.

flashback

Retrogression or burning back of the flame into or beyond the mixing chamber, sometimes accompanied by a hissing or squealing sound and the characteristic smoky, sharp pointed flame of small volume. When this occurs, immediately shut off the torch oxygen valve and then the acetylene valve.

flat position

Horizontal weld on the upper side of a horizontal surface.

flowability

Ability of a molten filler metal to flow or spread over a metal surface.

flux

Chemical compound or mixture in powdered, paste, or liquid form. Its essential function is to combine with, or otherwise render harmless, those products of the welding, brazing, or soldering operation that would reduce the physical properties of the deposited metal or make the welding, brazing, or soldering operation difficult or impossible.

forehand welding

Method of welding in which the torch and rod are so disposed in the vee. Torch flame points ahead in the direction of welding and the rod precedes the torch.

fuel gases

Gases usually used with oxygen for heating such as acetylene, natural gas, propane, methoacetylene, propadyne, and other synthetic fuels and hydrocarbons.

fuse plug

Safety device used on compressed gas cylinders with a low melting point alloy, designed to relieve excessive internal pressure due to heat by melting at a predetermined temperature.

fusion

Melting and flowing together of metals.

- G -

gas pocket

Cavity in a weld caused by entrapped fuel gas.

gas welding

Group of welding processes wherein fusion takes place as a direct result of the heat applied with a blowpipe fuel gas and oxygen.

generator

Apparatus for mechanically controlling the generation of acetylene by the reaction of calcium carbide and water.

gouging

Forming of a bevel or groove by removing material.

Groove

Opening provided by a grooved weld.

grooved weld

Welding rod fused into a joint that has the base metal removed to form a **V, U,** or **J** trough at the edge of the metals being joined.

- H -

hard facing or hard surfacing

Application of a hard, wear-resistant alloy to the surface of a softer metal by an arc or gas welding process.

heat

Molecular energy in motion.

heat-affected zone

Part of the base metal that has been altered by the heat from the welding, brazing, or cutting operation but that may not have actually melted.

heat conductivity

Speed and efficiency of heat energy movement through a substance.

horizontal position

Weld performed on a horizontal seam at least partially on a vertical surface.

hose

Flexible medium used to carry gases from the regulator to the torch. Made from rubber and reinforced with fabric.

hydrogen

Gas formed of the single element hydrogen. Considered one of the most active gases. When combined with oxygen, it forms a very clean flame. The flame, however, does not produce very much heat.

- I -

icicles

Undesirable condition wherein excess weld metal protrudes beyond the root of the weld or solder job.

inclusion

Gas bubble or nonmetallic particle entrapped in the weld metal as a result of improper torch flame or filler material manipulation.

incomplete fusion

Fusion that is less than complete.

inert gas

Gas that does not normally combine chemically with the base metal or filler metal.

infrared rays

Heat rays that come from both the arc and the welding flame.

inside-corner weld

Two metals fused together; one metal is held 90 degrees to the other. The fusion is performed inside the vertex of the angle.

intermittent weld

Joining two pieces and leaving un-welded sections in the joint.

- J -

joint

Where two pieces meet when a structure is made of smaller pieces.

joint design

Joint geometry together with the required dimensions of the welded joint.

joint penetration

Minimum depth a groove or flange weld extends from its face into a joint, exclusive of reinforcement.

-K-

kerf

Space from which metal has been removed by a cutting process.

keyhole

Enlarged root opening that is carried along ahead of the puddle when making an arc weld or other type of welded joint.

knee

Lower arm supporting structure in a resistance-welding machine.

-L-

land

Portion of the prepared edge of a part to be joined by a groove weld, which has not been beveled or grooved. Sometimes called *root face.*

lap joint

Welded joint in which two overlapping parts are connected, generally by means of fillet welds.

layer

Certain weld metal thickness made of one or more passes.

lens

Specially treated glass through which a welder may look at an intense flame without being injured by the harmful rays or glare radiating from the flame.

liquidation

Separation of a low melting constituent of an alloy from the remaining constituents, usually apparent in alloys having a wide melting range.

liquidus

Lowest temperature at which a metal or an alloy is completely liquid.

low-temperature brazing

Group of the brazing processes wherein the brazing alloys used melt in the range of about 1,175 to 1,300 degrees F and a shear (lap) joint is used.

-M-

malleable castings

Cast forms of metal that have been heat-treated to reduce their brittleness.

manifold

Multiple header for connection of individual gas cylinders or torch supply lines.

manual welding

Welding wherein the entire welding operation is performed and controlled by hand.

mapp

Stabilized methyl acetylene propadiene fuel gas often used in place of acetylene.

Martensite

A fully hardened steel which is very strong, but brittle unless further heat treated. The metal's structure is needle-like.

melting range

Temperature range between solidus and liquidus.

MIG

Gas metal arc welding (metal shielding gas).

mixing chamber

Part of the welding blowpipe where the welding gases are intimately mixed prior to release and combustion.

multilayer welding

Oxyacetylene welding, a technique in which a weld, on thick metal, is made in two or more passes.

-N-

neutral flame

A flame that results from combustion of perfect proportions of oxygen and the welding gas. Most commonly used flame for oxygen-fuel gas welding.

non-corrosive flux

Soldering flux that in itself and, as a residue, does not chemically attack the base metal. It is usually composed of rosin- or resin-base materials.

nonferrous

Metals containing no substantial amounts of ferrite or iron. These metals include copper, brass, bronze, aluminum, and lead.

normalizing

A method of controlling grain size after forging or working. The steel is heated above its upper critical temperature and cooled slowly in still air.

nozzle

See tip.

- O -

orifice

Opening, through which gases flow. It is usually the final opening or any opening controlled by a valve.

outside-corner weld

Fusing two pieces of metal together. Fusion takes place on the under part of the seam.

overhead position

Weld made on the underside of the joint with the face of the weld in a horizontal plane.

overlap

Extension of the weld face metal beyond the toe of the weld.

oxidation

Process of oxygen combining with elements to form oxides.

oxide

Chemical compound resulting from the combination of oxygen and other elements.

oxidizing flame

Flame produced by an excess of oxygen in the blowpipe mixture, leaving some free oxygen that tends to burn the molten metal.

oxygen

Gas formed of the element oxygen. When it is very actively it supports combustion, it is called burning; when it slowly combines with a substance, it is called oxidation, or rust.

oxyacetylene cutting

Cutting metal using the oxygen jet that is added to an oxygen-acetylene preheating flame.

oxyacetylene welding

Method of welding in which a fuel combination of two gases, oxygen and acetylene, is used.

oxygen cylinder

Specially built container used to store and/or transport oxygen.

oxygen hydrogen flame

Chemical combining of oxygen with the fuel gas hydrogen.

oxygen hose

See hose.

oxygen regulator

Automatic valve used to reduce cylinder pressures to torch pressures and to keep the working pressures constant.

-P-

parent metal

See base metal.

pass

Weld metal created by one progression along a weld.

peening

Mechanical working of metal by means of repeated hammer blows.

penetration

Penetration of a weld is the distance from the original surface of the base metal to that point at which fusion ceases.

plug weld

Weld that holds two pieces of metal together. It is made by making a hole in one piece of metal that is then lapped over the other piece.

porosity

Presence of gas pockets or voids in the metal or weld bead.

post-heating

Temperature to which a metal is heated after an operation has been performed on the metal such as welding, cutting, or forming.

preheating

The heating of a metal before an operation is performed on it such as welding, cutting, or forming.

psi

Standard abbreviation for pounds per square inch.

puddle

Portion of a weld that is molten at the place the heat is applied.

-Q-

quench

Cooling hot metal quickly by dunking in a liquid such as water or oil.

-R-

reducing flame

Oxygen fuel gas flame with a slight excess of the fuel gas.

regulator

Mechanical device for accurately controlling the pressure and flow of gases employed in welding, cutting, braze-welding, and so on.

reinforcement weld

Weld metal on the face of the weld in excess of that required for the size of the weld, for the purpose of adding strength.

resistance welding

Process using the resistance of the metals being welded to the flow of electricity as the source of the heat.

reversed polarity

Where the electrode is the positive anode, or reverse of most commonly used polarity.

root of weld

Part of a weld farthest from the application of weld heat and/or filler metal side.

-S-

safety disc

Mechanical safety device designed to release at a predetermined pressure.

skull

Un-melted residue from a liquefied filler metal.

slag inclusions

Non-fused, nonmetallic substances in the weld metal.

slugging

Act of adding a separate piece or pieces of material in a joint before or during welding, resulting in a welded joint that does not comply with the original design, drawing, or specification requirements.

soldering

Means of fastening metals together by adhering another metal to the two pieces of these metals. The joining metal only is melted during the operation. The joining metal melts below 800 degrees F.

solidus

Highest temperature at which a metal or alloy is completely solid.

spatter

Both arc and gas welding produce metal particles during the welding process that are not part of the weld.

spelter

Powdered brass used in making a typical brazed joint or lap joint.

spike welding

Type of resistance welding where large amounts of electricity are stored up and then rapidly released through the two metals to be joined. Spike welding has the advantage of being able to weld almost any kind of metal, alloy or dissimilar metals.

spot weld

Weld made between or upon overlapping members wherein fusion may start or occur on the faying (facing) surfaces or may have proceeded from the surface of one member. The weld cross section is approximately circular.

straight polarity*

Electrode is negative (−) or cathode. Connecting DC to cause electrons to flow in the direction desired.

---

* It should be noted that a debate still exists on the direction of electron flow in a circuit. The electronics technicians have been taught the electron flow and current flow both flow from negative to positive. The engineer has been taught that current flow is positive to negative. So, it up to the reader to determine for himself which direction the current flows. In dealing with semiconductors it is argued that current flow is in one direction and electron flow is another for the sake of explaining how a transistor works in a circuit.

strain

Reaction of an object to stress.

stress

Load imposed on an object.

stress relieving

Even heating of a structure to a temperature below the critical temperature followed by a slow, even cooling.

surfacing

Deposition of a filler metal on a metal surface to obtain desired properties or dimensions.

sweat soldering

Soldering method in which two or more parts that have been pre-coated with solder are reheated and assembled into a joint without the use of additional solder.

-T-

tack weld

Small weld used to temporarily hold together components of an assembly until they can be welded.

tank

See cylinder.

T-joint

Joint formed by placing one metal against another at an angle of 90 degrees. The edge of one metal contacts the surface of the other metal.

tempering

the reduction of hardness of steels by controlled heating between 300 degrees F and 750 degrees F.

tensile strength

Maximum pull stress, in pounds per square inch, that a specimen is capable of developing.

thermit welding

Group of welding processes that achieve coalescence by pouring a superheated liquid metal (thermit) around the parts to be welded. Thermit is a trade name for a mixture of finely divided metal oxide and a metal reducing agent.

throat of fillet weld

Distance from weld face to weld root.

TIG

Tungsten inert gas welding.

tinning

Coating of the metals to be soldered with soldering metal.

tip

Part of the torch at the end where the gas exits and burns. In arc welding it is the end of the electrode.

toe of weld

Junction of the face of the weld and the base metal.

torch

Mechanism that the operator holds during gas welding and cutting, at the end of which the gases are burned to perform various gas welding and cutting operations. Often called the *blowpipe*.

-U-

ultraviolet rays

Energy waves that emanate from the electrodes and the welding flames at a frequency in the ultraviolet ray light spectrum.

undercut

Depression at the toe of the weld that is below the surface of the metal.

underfill

Depression on the face of the weld or root surface extending below the surface of the adjacent base meta.

-V-

vee groove

See butt joint.

vertical position

Type of weld where the welding is done on a vertical seam and surface.

voltage regulator

Automatic electrical control device for maintaining a constant voltage supply to the welding transformer.

- W -

welding

Art of fastening metals together by means of inter-fusing the metals.

welding rod

Wire that is melted into the weld metal.

welding sequence

Order in which the component parts of a structure are welded.

weldment

Order in which the components parts of a structure are welded.

weld metal

Fused portion of base metal or fused portion of both the base metal and the filler metal.

weld pool

Small body of molten metal created by the flame of the torch.

work hardening

Increase in strength and hardness produced by working certain metals such as iron, copper, aluminium, and nickel. It is most pronounced in cold welding.

- Y -

yield strength

Stress, in pounds per square inch, at which a specimen assumes a specified limiting permanent set.

- Z -

zirconium

Extremely hard metal. Used as an alloy material in some instances. A silvery-white metal that is available in sheets, foil, wire and rod. Used as an alloy to add thermal stability and corrosion resistance.

—NOTES—

Appendix 5
REFERENCE MATERIAL

Welding Licensing Exam Study Guide

WELDING NOZZLE FLOW DATA

Metal Thickness	Tip Size	Drill Size	Oxygen Pressure (PSIG)		Acetylene Pressure (PSIG)		Acetylene Consumption (SCFH)	
			Min.	Max.	Min.	Max.	Min.	Max.
Up to 1/32"	000	75 (.022)	3	5	3	5	1	2
1/16"-3/64"	00	70 (.028)	3	5	3	5	1½	3
1/32"-5/64"	0	65 (.035)	3	5	3	5	2	4
3/64"-3/32"	1	60 (.040)	3	5	3	5	3	6
1/16"-1/8"	2	56 (.046)	3	5	3	5	5	10
1/8"-3/16"	3	53 (.060)	4	7	3	6	8	18
3/16"-1/4"	4	49 (.073)	5	10	4	7	10	25
1/4"-1/2"	5	43 (.089)	6	12	5	8	15	35
1/2"-3/4"	6	36 (.106)	7	14	6	9	25	45
3/4"-1¼"	7	30 (.128)	8	16	8	10	30	60
1¼"-2"	8	29 (.136)	10	19	9	12	35	75
2½"-3"	10	27 (.144)	12	24	12	15	50	100
3½"-4"	12	25 (.149)	18	28	12	15	80	160

TYPE MFA HEATING NOZZLES

TIP SIZE	ACETYLENE PRESSURE RANGE P.S.I.G.	OXYGEN PRESSURE RANGE P.S.I.G.	ACETYLENE CUBIC FEET PER HOUR		OXYGEN CUBIC FEET PER HOUR		BTU PER HOUR
			Minimum	Maximum	Minimum	Maximum	
4	6-10	8-12	6	20	7	22	
6	8-12	10-15	14	40	15	44	
8	10-15	20-30	30	80	33	88	SEE NOTES
10	12-15	30-40	40	100	44	110	
12*	12-15	50-60	60	150	66	165	
15*	12-15	50-60	90	220	99	244	

WARNING: At no time should the withdrawal rate of an individual acetylene cylinder exceed 1/7 of the cylinder contents per hour. If additional flow capacity is required, use an acetylene manifold system of sufficient size to supply the necessary volume.

MFN HEATING NOZZLES

Tip Size	Propane Pressure PSIG	Oxygen Pressure PSIG	Propane Cubic Feet Per Hour		Oxygen Cubic Feet Per Hour		BTU Per Hour
			Minimum	Maximum	Minimum	Maximum	
6	2	8	5.5	20	20	80	
8	2	10-20	10	35	40	140	
10	4	10-30	20	80	80	320	SEE
12*	5-25	30-125	30	160	120	640	BELOW
15*	5-25	30-125	50	200	200	800	
20*	5-30	40-135	75	250	300	1000	

*Use model HD310C torch and ⅜″ hose.

TYPE 55 NOZZLES

TIP SIZE	OXYGEN PRESSURE (psig)	FUEL GAS PRESSURE (psig)	CONSUMPTION (SCFH)		BTU PER HOUR
			OXYGEN	FUEL GAS	
10*	70/100	15/25	350/460	150/200	SEE
15*	90/120	20/35	600/800	250/350	NOTES
20*	100/150	30/50	900/1150	400/500	

* Use Model HD310C torch and ⅜″ hose.

NOTE: Approximate gross BTU contents per cubic foot:
- Acetylene-1470
- Propane-2458
- Methane-1000
- Butane-3374
- MAPP-2406
- Propylene-2371
- Nat. Gas-1000

TYPES 1-101, 3-101 & 5-101 (Oxy-Acetylene)

Metal Thickness	Tip Size	Cutting Oxygen		Pre-heat Oxygen* (PSIG)	Acetylene		Speed I.P.M.	Kerf Width
		Pressure*** (PSIG)	Flow**** (SCFH)		Pressure (PSIG)	Flow (SCFH)		
1/8"	000	20/25	20/25	3/5	3/5	6/11	20/30	.04
1/4"	00	20/25	30/35	3/5	3/5	6/11	20/28	.05
3/8"	0	25/30	55/60	3/5	3/5	6/11	18/26	.06
1/2"	0	30/35	60/65	3/6	3/5	9/16	16/22	.06
3/4"	1	30/35	80/85	4/7	3/5	8/13	15/20	.07
1"	2	35/40	140/160	4/8	3/6	10/18	13/18	.09
2"	3	40/45	210/240	5/10	4/8	14/24	10/12	.11
3"	4	40/50	280/320	5/10	5/11	18/28	10/12	.12
4"	5	45/55	390/450	6/12	6/13	22/30	6/9	.15
6"	6**	45/55	500/600	6/15	8/14	25/35	4/7	.15
10"	7**	45/55	700/850	6/20	10/15	25/35	3/5	.34
12"	8**	45/55	900/1050	7/25	10/15	25/35	3/4	.41

*Applicable for 3-hose machine cutting torches only. With a two hose cutting torch, preheat pressure is set by the cutting oxygen.

**For best results use ST 1600C-ST 1900C series torches and 3/8" hose using tip size 6 and larger.

***All pressures are measured at the regulator using 25'x1/4" hose through tip size 5, and 25'x3/8" hose for tip size 6 and larger.

WARNING: At no time should the withdrawal rate of an individual acetylene cylinder exceed 1/7 of the cylinder contents per hour. If additional flow capacity is required, use an acetylene manifold system of sufficient size to supply the necessary volume.

TYPES 303M, GPM, GPN, GPP

Metal Thickness	Tip Size	Cutting Oxygen		Pre-heat Oxygen (PSIG)	Pre-heat Fuel Gas**		Speed I.P.M.	Kerf Width
		Pressure (PSIG)	Flow (SCFH)		Pressure (PSIG)	Flow (SCFH)		
⅛"	000	20/25	12/14		3/5	5/6	20/30	.04
¼"	00	20/25	22/26		3/5	5/7	20/28	.05
⅜"	0	25/30	45/55		3/5	8/10	18/26	.06
½"	0	30/35	50/55	S	3/5	8/10	16/22	.06
¾"	1	30/35	70/80	E	4/6	10/12	15/20	.08
1"	2	35/40	115/125	E	4/8	12/15	13/20	.09
1½"	2	40/45	125/135		4/8	12/15	13/18	.09
2"	3	40/45	150/175	N	5/9	14/18	11/13	.10
2½"	3	45/50	175/200	O	5/9	14/18	10/12	.10
3"	4	40/50	210/250	T	6/10	16/20	8/10	.12
4"	5	45/55	300/360	E	8/12	20/30	6/9	.14
5"	5	50/55	330/360	S	8/12	20/30	4/7	.14
6"	6	45/55	400/500		10/15	25/35	3/5	.17
8"	6	55/65	450/500		10/15	25/35	3/4	.18
12"	8**	60/70	750/850		10/14	25/120	3/4	.41

*NOTE: The above data applies to all torches with the following exceptions:

Torch Series	Pre-Heat Oxygen	Pre-heat Fuel
MT 600N Series	25 PSIG-UP	8 OZ.-UP
ST 900C Series	N/A	5 PSIG-UP
ST 1600C/1700C Series	N/A	1 PSIG-UP
ST 1800C/1900C Series	N/A	1 PSIG-UP
MT 200 Series	N/A	8 OZ-UP
MT 300 Series	10-50 PSIG	8 OZ.-UP

**For best results use ST 1600C — ST 1900C series torches and ⅜" hose when using tip size 6 or larger.

***All pressures are measured at the regulator using 25'x⅜" hose for tip size 6 and larger.

WARNING: High gas withdrawal rates require use of a manifold system of sufficient size to supply the necessary volume.

TYPES: MTHM, P

Metal Thickness	Tip Size	Cutting Oxygen Pressure*** (PSIG)	Flow (SCFH)	Pre-Heat Oxygen* Pressure (PSIG)	Flow (SCFH)	Pre-Heat Fuel Gas Pressure (PSIG)	Flow (SCFH)	Speed I.P.M.	KERF Width
1/4"	00	85/95	68/75		23/140		12/65	23/30	.05
3/8"	00	85/95	68/75		23/140		12/65	22/29	.05
1/2"	0	85/95	110/120		23/140		12/65	20/28	.06
3/4"	0	85/95	110/120		23/140		12/65	18/26	.06
1"	1	85/95	145/160	S	23/140	S	12/65	17/24	.07
1 1/4"	1	85/95	145/160	E	23/140	E	12/65	16/20	.07
1 1/2"	1	85/95	145/160	E	23/140	E	12/65	12/16	.07
2"	2	85/95	230/250		23/140		12/65	11/15	.09
2 1/2"	2	85/95	230/250	N	23/140	N	12/65	10/13	.09
3"	2	85/95	230/250	O	23/140	O	12/65	9/11	.09
4"	3	85/95	285/320	T	23/140	T	12/65	7/10	.11
5"	3	85/95	285/320	E	23/140	E	12/65	6/8	.11
6"	3	85/95	285/320	S	23/140	S	12/65	5/7	.11
7"	4	85/95	390/450		23/140		12/65	5/6	.14
8"	4	85/95	390/450		23/140		12/65	4/6	.14
9"	5	85/95	670/720		23/140		12/65	4/5	.18
10"	5	85/95	670/720		23/140		12/65	3/5	.18

*NOTE: The above data applies to all torches with the following exceptions:

Torch Series	Pre-Heat Oxygen	Pre-Heat Fuel
MT 300N Series	10-50 PSIG	8 Oz.-UP
MT 600N Series	25 PSIG-UP	8 Oz.-UP

***All pressures are measured at the torch inlet on MTH Series Tips.

WARNING: High gas withdrawal rates require use of a manifold system of sufficient size to supply the necessary volume.

TYPE: HPM, N, P

Metal Thickness	Tip Size	Cutting Oxygen		Pre-Heat Oxygen		Pre-Heat Fuel Gas		Speed I.P.M.	KERF Width
		Pressure*** (PSIG)	Flow (SCFH)	Pressure (PSIG)	Flow* (SCFH)	Pressure* (PSIG)	Flow (SCFH)		
3/4"	1	$^{30}/_{35}$	$^{70}/_{80}$	S	$^{44}/_{240}$	$^{3}/_{6}$	$^{22}/_{110}$	$^{15}/_{20}$	.08
1"	2	$^{35}/_{40}$	$^{115}/_{125}$	E	$^{44}/_{240}$	$^{3}/_{6}$	$^{22}/_{110}$	$^{14}/_{18}$	.09
1½"	2	$^{40}/_{45}$	$^{125}/_{135}$	E	$^{44}/_{240}$	$^{4}/_{8}$	$^{22}/_{110}$	$^{12}/_{16}$	.09
2"	3	$^{40}/_{45}$	$^{150}/_{175}$		$^{44}/_{240}$	$^{4}/_{8}$	$^{22}/_{110}$	$^{10}/_{14}$	.10
2½"	3	$^{45}/_{50}$	$^{175}/_{200}$		$^{44}/_{240}$	$^{5}/_{9}$	$^{22}/_{110}$	$^{9}/_{12}$	.10
3"	4	$^{40}/_{50}$	$^{210}/_{250}$	N	$^{44}/_{240}$	$^{6}/_{9}$	$^{22}/_{110}$	$^{8}/_{11}$	.12
4"	5	$^{45}/_{55}$	$^{300}/_{360}$	O	$^{44}/_{240}$	$^{6}/_{9}$	$^{22}/_{110}$	$^{7}/_{10}$	.14
5"	5	$^{50}/_{55}$	$^{330}/_{360}$	T	$^{44}/_{240}$	$^{6}/_{10}$	$^{22}/_{110}$	$^{6}/_{9}$	.14
6"	6**	$^{45}/_{55}$	$^{400}/_{500}$	E	$^{44}/_{240}$	$^{6}/_{10}$	$^{22}/_{110}$	$^{5}/_{7}$	.17
8"	6**	$^{55}/_{65}$	$^{450}/_{500}$	S	$^{44}/_{240}$	$^{8}/_{12}$	$^{22}/_{110}$	$^{4}/_{6}$	.18
12"	8**	$^{60}/_{70}$	$^{750}/_{850}$		$^{50}/_{265}$	$^{10}/_{14}$	$^{25}/_{120}$	3/4	.41
15"	10**	$^{45}/_{55}$	$^{1000}/_{1200}†$		$^{50}/_{265}$	$^{10}/_{16}$	$^{25}/_{120}$	$^{2}/_{4}$	—
18"	12**	$^{45}/_{55}$	$^{1150}/_{1350}†$		$^{60}/_{290}$	—	$^{30}/_{130}$	$^{2}/_{3}$	—

*Applicable for 3-hose machine cutting torches only. With a two hose cutting torch, preheat pressure is set by the cutting oxygen.

**For best results use ST 1600C-ST 1900C series torches and ⅜" hose using tip size 6 and larger.

***All pressures are measured at the regulator using 25'x ¼" hose through tip size 5, and 25"x⅜ hose for tip size 6 and larger.

WARNING: High gas withdrawal rates require use of a manifold system of sufficient size to supply the necessary volume.

NOTE: The above data applies to all torches with the following exceptions:

Torch Series	Pre-Heat Oxygen	Pre-Heat Fuel
MT 200N Series	N/A	8 oz.-UP
MT 300N Series	10-50 PSIG	8 oz.-UP
MT 600N Series	25 PSIG-UP	8 oz.-UP
ST 900C Series	N/A	5 PSIG-UP
ST 1600C-ST 1900C	N/A	2 PSIG-UP
STN 2300 Series	N/A	2 PSIG-UP

CUTTING TIP PREHEAT AND CUTTING ORIFICE DRILL SIZE

Tip Size Orifice Size*	Cutting Oxygen Size	Cleaning Drill	1-100 3-100	1-101 3-101 5-101	1-104	1-108 3-108	1-110 3-110 5-110	1-111	1-112 3-112	1-129	1-200 3-200 5-200	1-116 3-116	1-117	1-118 3-118	1-207	1-218
000	71	72		74												
00	67	68		74			65		67							
0	60	61	71	74		75	60	64	60		67			71		
1	56	57	67	71		73	56		56		64					
2	53	54	60	67		66	54	56	53	57	62	66		63	55	
3	50	51		66		63	53		52		60	64				
4	45	46		66		60		53	52	55	56	61		56	53	57
5	39	40		66							55					
6	31	32		63							54			57		57
7	28	29		63												
8	20	21		63	55								63	57		56
10	13	14			55									57		55
12	2	$\frac{7}{32}$												56		

Same for all types except scarfing, gouging, deseaming and high speed.

Appendix 6
SMAW CARBON STEEL ELECTRODES

Welding Licensing Exam Study Guide

The mechanical properties and chemical analyses listed for these electrodes are nominal ones. They do vary from one manufacturer to the next for a particular electrode. Whenever possible, always check the manufacturer's electrode specifications.

TABLE A6.1   Heavy coated mild steel electrodes.

Electrode Type	Comments
E6010	• <u>General</u>: an all-purpose, fast freeze electrode used to weld carbon and galvanized steel. It can be used for welding through light to medium amounts of dirty, rusty or painted materials on surfaces that cannot be completely cleaned. The E6010 is recommended for welding API grades A25, A, B, and X42 pipe; for welding out-of-position X-ray welds; for general purpose and maintenance welding; for welding pressure vessel fittings; for construction and shipbulding welding, and storage tank fabrication. • <u>Electrode coating</u>: Organic coating. High cellulose sodium, 0-10% iron powder. • <u>Welding current/polarity</u>: Direct Current, electrode positive (DCEP) • <u>Arc type and penetration</u>: quick starting, deep penetrating, steady spray type arc with average deposition rate. • <u>Welding positions:</u> All. Good metal transfer and rapid set up in vertical up, vertical down, and overhead welding positions. • <u>Welding characteristics</u>: moderate spatter and moderately easy slag removal. Weld puddle wets and spreads well. Weld bead has rippled and flat appearance. • <u>Typical mechanical properties</u>: Minimum tensile strength: 70,000-79,800 psi; yield point: 61,000 psi; minimum elongation in 2": 22-29%; charpy V-notch (@-20°F) 36 ft. lbs. • <u>Typical chemical analysis</u>: Carbon (0.13%), Manganese (0.51%), Phosphorous (0.014%), Sulfur (0.006%), Silicon (0.20%).
E6011	• <u>General</u>: an all-purpose, fast-freeze electrode used to weld carbon and galvanized steel and for rusty and oily steel in maintenance and repair work. It has the same characteristics as the E6010, electrode but can be used with AC and DC currents. Typical applications include general purpose repair and fabrication work; galvanized steel work; shipbuilding; structural work; welding truck frames, storage tanks, railway cars, heavy equipment, and boilers. Commonly used to weld mild steels such as ASTM A-36, A-283, A-284, A-285, A-515, and A-516. • <u>Electrode coating</u>: Organic coating. High cellulose potassium, 0% iron powder. • <u>Welding current/polarity</u>: AC current. The E6011 electrode can also be used with DCEP or DCEN current, but with decreased joint penetration. AC current eliminates the arc blow problem sometimes encountered with the E6010 electrode running on DC current. • <u>Arc type and penetration</u>: Strong arc with deep penetration and fine spray transfer.

Electrode Type	Comments
E6011 *(continued)*	• <u>Welding Positions</u>: All positions. Fast freezing (rapid solidification) of the weld metal allows welding in the vertical and overhead positions • <u>Welding Characteristics</u>: Moderate spatter, but more spatter than the E6010 electrode. Moderately easy slag removal. Note: light slag tends to eliminate slag holes. Average deposition rate. The weld deposit is free from porosity, holes, and pits. Rippled and flat bead appearance. Fillet and bead contours are flat instead of convex. • <u>Typical mechanical properties</u>: Similar to E6010 electrode. • <u>Typical chemical analysis: </u>: Similar to E6010 electrode.
E6012	• <u>General</u>: Fill-freeze type electrode used in general repair and fabrication work; galvanized steel work; shipbuilding; structural work; welding truck frames, storage tanks, railway cars, heavy equipment, and boilers. Widely used in most steel fabrication. Recommended in applications with joint fit up. E6012 capable of filling wide joint gaps. Economical to use because of its rapid welding speed and ease of use. • <u>Electrode Coating:</u> Rutile coating. High titania sodium, 0-10% iron powder. • <u>Welding current/polarity</u>: AC or DCEN. Use DCEN in all positions except when arc below becomes a problem. • <u>Arc type and penetration</u>: Medium arc with medium penetration. Above average deposition rate. • <u>Welding positions:</u> All positions. Most commonly used in downhill or flat (horizontal) position. • <u>Welding characteristics</u>: Slight spatter. Produces light but easily removable slag at fast speeds. More slag than the E6010 electrode, but less than the E6020. Produces small, smooth, and convex beads. • <u>Typical mechanical properties</u>: Minimum tensile strength: 67,000-69,400 psi; yield point 55,000-62,300 psi: minimum elongation in 2": 17-21% • <u>Typical chemical analysis</u>: Carbon (0.05%), Manganese (0.31%), phosphorous (0.008%), sulfur (0.016%), silicon (0.12%)
E6013	• <u>General</u>: Fill-freeze type electrode. General purpose stick electrode for use on carbon steels. Wide variety of applications for light gauge and heavy plate. Recommended for applications with poor joint fit-up. Commonly used for farm equipment, metal furniture, auto bodies, machine parts, shaft buildup, and other applications where low spatter and good bead appearance is desirable. • <u>Electrode Coating:</u> Rutile coating. High titania sodium, 0-10% iron powder. • <u>Welding current/polarity</u>: AC or DCEN. Use DCEN in all positions except when arc blow becomes a problem. Works well on low voltage AC machines. • <u>Arc type and penetration</u>: Soft arc with light to medium penetration. Above average deposition rate. • <u>Welding positions:</u> All positions. Recommended for vertical down welding. • <u>Welding characteristics</u>: Spatter is exceedingly low. Slag is easily removed and often self cleaning on heavy plate. E6013 electrode produces a smooth and flat to convex weld bead with fine ripple appearance. • <u>Typical mechanical properties</u>: Minimum tensile strength: 67,000-71,000 psi; yield point: 62,000-63,000 psi; minimum elongation in 2": 17-24%; charpy V-notch (@32°F) 58 ft. lbs.

(continued on next page)

Electrode Type	Comments
E7014	• <u>General</u>: Recommended for welding machine bases, frames, heavy sheet metal, and for general repair and maintenance work. Often recommended for applications where poor fit-up is encountered or a higher deposition rate and travel speed is required. • <u>Electrode Coating</u>: Rutile iron powder coating. • <u>Welding current/polarity</u>: AC, DCEN. DCEP • <u>Arc type and penetration</u>: AC, DCEN and DCEP. • <u>Welding positions</u>: All positions. • <u>Welding characteristics</u>: Easily removable slag. Produces a smooth bead surface with fine ripples. • <u>Typical mechanical properties</u>: Minimum tensile strength: Up t0 70,000 psi; yield point: 67,700 psi; minimum elongation (in 2") 29.4%. • <u>Typical chemical analysis</u>: Carbon (0.12%), Chromium (0.041%), Copper (0.012%), Manganese (0.68%), Phosphorous (0.021%), Vanadium (0.023%), Molybdenum (0.002%), Nickel (0.053%), Sulfur (0.012%), Silicon (0.33%), Iron (balance).
E7015	• <u>General</u>: A low-hydrogen electrode. Recommended for welding high carbon, high sulfur, and alloy steels. • <u>Electrode Coating:</u> Low hydrogen sodium • <u>Welding current/polarity</u>: DCEP • <u>Arc type and penetration</u>: Mild to medium penetration. Welding with a short arc is important for obtaining high quality weld deposits. Above average deposition rate. • <u>Welding positions</u>: All positions. • <u>Welding characteristics</u>: Slight spatter. Heavy, but easily removable slag. E7015 electrode produces a flat and slightly convex weld bead . • <u>Typical mechanical properties</u>: Minimum tensile strength: 70,000 psi; yield point: 60,000 psi; minimum elongation (in 2") 22%. • <u>Typical chemical analysis</u>: Carbon (0.08-0.13%), Manganese (0.40-0.60%), phosphorous (0.04%), sulfur (0.04%), silicon (0.25%)
E7016	• <u>General</u>: Low hydrogen electrode. Recommended for welding high carbon, low alloy, sulfur • <u>Electrode Coating:</u> Low hydrogen potassium. • <u>Welding current/polarity</u>: DCEP, AC • <u>Arc type and penetration</u>: Mild to medium penetration. Welding with a short arc is important for obtaining high quality weld deposits. • <u>Welding positions</u>: All positions. • <u>Welding characteristics</u>: Slight spatter with very easy slag removal. The E7016 electrode produces a smooth and convex weld bead. Above average deposition rate. • <u>Typical mechanical properties</u>: Minimum tensile strength: 70,000-79,000 psi; yield point: 60,000-69,500 psi; minimum elongation in 2": 22-28%. • <u>Typical chemical analysis</u>: Carbon (0.07%), Manganese (0.09%); Silicon (0.50%)
E7018	• <u>General</u>: A low-hydrogen, iron-powder type, fill-freeze electrode used to weld low, medium and high-carbon steels. Recommended for joints involving high-strength and high carbon. Commonly used for pipe, heavy sections of plate boiler work, and low temperature equipment. • <u>Electrode Coating</u>: A low hydrogen base flux with iron powder added. • <u>Welding current/polarity</u>: DCEP or AC. Use AC current with electrode diameters larger then 5/32". Use DCEP with diameters 5/32" and smaller. Note: Not recommended for low-voltage AC welders. • <u>Arc type and penetration</u>: Smooth, quiet arc and medium arc penetration. Note: As is common with all low hydrogen electrodes, a short arc length should be maintained at all times. E7018 type electrodes require a "drag" to 1/16" maximum arc length to obtain the

Electrode Type	Comments
E7018 *(continued)*	desired weld quality and mechanical properties. Porosity and a deterioration of impact properties may occur wit an arc length of 1/8" or longer. • Welding positions: All positions. E7018 electrode is often recommended for out-of-position welding and tacking. Note: Fillet welds in the horizontal and flat welding positions have a slightly convex weld face. • Welding characteristics: Slight spatter with very easy slag removal. Flat to convex bead appearance with a smooth and finely rippled bead surface. The E7018 electrode is very susceptible to moisture, which may lead to weld porosity. Sometimes used as the final bead or layer (for better appearance) after laying the first bead with an E6010 electrode. Medium to high deposition rate. • Typical mechanical properties: Minimum tensile strength: 72,000-78,000 psi; yield point: 60,000-68,000 psi; minimum elongation in 2": 22-31%; Charpy V-notch (@-20ºF) 65 ft. lbs. • Typical chemical analysis: Carbon (0.04%), Manganese (1.06%), phosphorous (0.012%), sulfur (0.011%), silicon (0.69%).
E7028	• General: Low hydrogen electrode, fast-fill type electrode used in high production, fast deposition rate welding. • Electrode Coating: A low hydrogen base flux with iron powder added. Coating contains a higher percentage of iron powder (50%) than the E7018 electrode. Thicker and heavier coating than the E7018. • Welding current/polarity: AC or DCEP • Arc type and penetration: Shallow to Mild penetration. Welding with a short arc is recommended for obtaining high quality weld deposits. • Welding positions: Horizontal and flat positions. Recommended for flat fillets and deep groove joints. • Welding characteristics: Slight spatter with very easy slag removal. Heavy slag peals off. Smooth and slightly convex bead appearance. Excellent restrikng for skip and tack welding. Fastest deposition rate among all the electrodes. • Typical mechanical properties: Minimum tensile strength: 72,000 psi; yield point: 60,000 psi; minimum elongation in 2": 22%. • Typical chemical analysis: Carbon (0.04%), Manganese (1.06%), phosphorous (0.012%), sulfur (0.011%), silicon (0.69%).
E7024	• General: Iron powder, high speed, heavy coated electrode. Typical applications include shipbuilding, bridge construction, structural steels, storage tanks, and truck frames. • Electrode Coating: Iron powder titania. Rutile type with higher efficiency (more iron powder) than type E7014. • Welding current/polarity: AC, DCEP, and DCEN • Arc type and penetration: Mild penetration. Increased penetration with little or no root porosity in horizontal or positioned fillets. • Welding positions: Suitable for use in horizontal (flat) position and for standing fillets. • Welding characteristics: Slight spatter. Easy slag removal. Often self cleaning. Produces smooth and convex bead appearance. Very high deposition rate. • Typical mechanical properties: Minimum tensile strength: 72,000-81,000 psi; yield point: 60,000- 71,000 psi: minimum elongation (in 2")17-26%. • Typical chemical analysis: Carbon (0.06%), Manganese (0.81%), phosphorous (0.018%), sulfur (0.019%), silicon (0.43%)

(continued on next page)

Electrode Type	Comments
E6027	• <u>General</u>: Iron powder electrode recommended where high quality groove and fillet welds are required. Suitable for heavy structural welding. • <u>Electrode Coating</u>: Iron powder iron oxide. • <u>Welding current/polarity</u>: AC, DCEN, and DCEP • <u>Arc type and penetration</u>: Medium arc with medium penetration. • <u>Welding positions</u>: Horizontal fillet welds and flat position. • <u>Welding characteristics</u>: Slight spatter with easy slag removal. High deposition rate, but slower than the E7024 electrode. Weld properties of E6027 superior to those of E7024 electrode. Produces a flat to concave bead. • <u>Typical mechanical properties</u>: Minimum tensile strength: 62,000-72,000 psi; yield point: 48,000-60,000 psi; minimum elongation (in 2") 25%. • <u>Typical chemical analysis</u>: Carbon (0.05-0.10%), Manganese (0.65-0.95%), phosphorous (0.015-.025%), sulfur (0.10-0.025%), silicon (0.010-.025%)